Unit 1
Plant Shapes

Copyright © 2020 by Discovery Education, Inc. All rights reserved. No part of this work may be reproduced, distributed, or transmitted in any form or by any means, or stored in a retrieval or database system, without the prior written permission of Discovery Education, Inc.

NGSS is a registered trademark of Achieve. Neither Achieve nor the lead states and partners that developed the Next Generation Science Standards were involved in the production of this product, and do not endorse it.

To obtain permission(s) or for inquiries, submit a request to:
Discovery Education, Inc.
4350 Congress Street, Suite 700
Charlotte, NC 28209
800-323-9084
Education_Info@DiscoveryEd.com

ISBN 13: 978-1-68220-787-1

Printed in the United States of America.

5 6 7 8 9 10 CWM 26 25 24 23   B

Acknowledgments

Acknowledgment is given to photographers, artists, and agents for permission to feature their copyrighted material.

Cover and inside cover art: Rawpixel.com / Shutterstock.com

Table of Contents

Unit 1

Letter to the Parent/Guardian v

Plant Shapes viii

   Get Started: Need for Urban Gardens 2

Unit Project Preview: Pizza Garden 4

Concept 1.1

Plants for a Purpose 6

   Wonder 8

      Let's Investigate Garden Salads 10

   Learn 20

   Share 48

Concept 1.2

Growing Plants 58

   Wonder 60

      Let's Investigate Hydroponic Plants 62

   Learn 74

   Share 100

Unit 1: Plant Shapes

Concept 1.3
Designing for Plants 110

   Wonder 112

      Let's Investigate Creepers 114

   Learn 124

   Share 158

Unit Project
Unit Project: Pizza Garden 168

Grade 1 Resources
Bubble Map R1

Safety in the Science Classroom R3

Vocabulary Flash Cards R7

Glossary R17

Index R29

Dear Parent/Guardian,

This year, your student will be using Science Techbook™, a comprehensive science program developed by the educators and designers at Discovery Education and written to the Next Generation Science Standards (NGSS). The NGSS expect students to act and think like scientists and engineers, to ask questions about the world around them, and to solve real-world problems through the application of critical thinking across the domains of science (Life Science, Earth and Space Science, Physical Science).

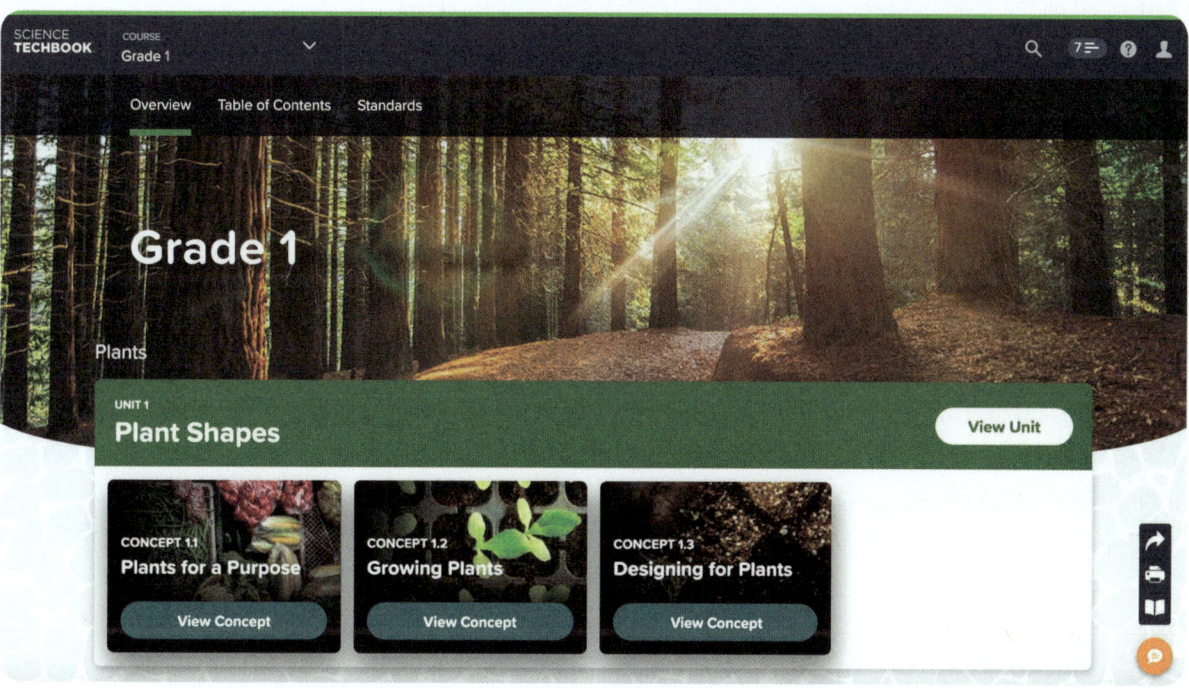

Science Techbook is an innovative program that helps your student master key scientific concepts. Students engage with interactive science materials to analyze and interpret data, think critically, solve problems, and make connections across science disciplines. Science Techbook includes dynamic content, videos, digital tools, Hands-On Activities and labs, and gamelike activities that inspire and motivate scientific learning and curiosity.

You and your child can access the resource by signing in to www.discoveryeducation.com. You can view your child's progress in the course by selecting Assignments.

Science Techbook is divided into units, and each unit is divided into concepts. Each concept has three sections: Wonder, Learn, and Share.

Units and Concepts Students begin to consider the connections across fields of science to understand, analyze, and describe real-world phenomena.

Wonder Students activate their prior knowledge of a concept's essential ideas and begin making connections to a real-world phenomenon and the **Can You Explain?** question.

Learn Students dive deeper into how real-world science phenomenon works through critical reading of the Core Interactive Text. Students also build their learning through Hands-On Activities and interactives focused on the learning goals.

Share Students share their learning with their teacher and classmates using evidence they have gathered and analyzed during Learn. Students connect their learning with STEM careers and problem-solving skills.

vi

Within this Student Edition, you'll find QR codes and quick codes that take you and your student to a corresponding section of Science Techbook online. To use the QR codes, you'll need to download a free QR reader. Readers are available for phones, tablets, laptops, desktops, and other devices. Most use the device's camera, but there are some that scan documents that are on your screen.

For resources in Science Techbook, you'll need to sign in with your student's username and password the first time you access a QR code. After that, you won't need to sign in again, unless you log out or remain inactive for too long.

We encourage you to support your student in using the print and online interactive materials in Science Techbook on any device. Together, may you and your student enjoy a fantastic year of science!

Sincerely,

The Discovery Education Science Team

Unit 1: Plant Shapes

Unit 1
Plant Shapes

Get Started

Need for Urban Gardens

Food deserts are places where people cannot find fresh food. In this unit, you will learn about different types of plants and how they grow. At the end of the unit, you will be able to use the information you learned to design an urban garden to grow fresh food.

How can people in cities get fresh food to eat?

Watch a video about solutions for growing gardens in cities.

Quick Code: us1006s

Need for Urban Gardens

Think About It

Look at the picture. Think about the following questions:

- How can we tell that plants are different from one another?

- How do these differences help the plants?

Food in an Urban Garden

Unit 1: Plant Shapes | 3

Unit Project Preview

Design Solutions Like a Scientist

Quick Code: us1007s

Hands-On Engineering: Pizza Garden

In this activity, you will design and plan a garden that can grow toppings for pizza.

Tomato Plant

- **SEP** Asking Questions and Defining Problems
- **SEP** Developing and Using Models
- **SEP** Constructing Explanations and Designing Solutions
- **SEP** Analyzing and Interpreting Data
- **CCC** Scale, Proportion, and Quantity
- **CCC** Structure and Function

Ask Questions About the Problem

You are going to design a garden that has fruits and vegetables you can put on a pizza. **Write** some questions you can ask to learn more about the problem. As you work on activities throughout the unit, **write** down answers to your questions.

Unit 1: Plant Shapes | 5

CONCEPT
1.1

Plants for a Purpose

Student Objectives

By the end of this lesson:

- ☐ I can make observations and contrast the parts of different plants.
- ☐ I can find patterns when comparing young plants and adult plants.

Key Vocabulary

- ☐ edible
- ☐ energy
- ☐ engineer
- ☐ flower
- ☐ fruit
- ☐ plant
- ☐ seed
- ☐ stem
- ☐ trait

Quick Code: us1009s

1.1 Plants for a Purpose

Activity 1

Can You Explain?

How do different plants change over time?

Quick Code:
us1010s

1.1 Plants for a Purpose

1.1 | Wonder How do different plants change over time?

Activity 2
Ask Questions Like a Scientist

Quick Code: us1011s

Garden Salads

Look at the photo. **Answer** the questions.

Let's Investigate Garden Salads

SEP Asking Questions and Defining Problems

10

What fruits and vegetables do you see?

Think about how the parts of the salad grew. What questions do you have about how plants change over time?

Your Questions

1.1 Plants for a Purpose

1.1 | Wonder How do different plants change over time?

Activity 3

Observe Like a Scientist

Quick Code: us1012s

Red Peppers

Look at the image. **Talk** about the question with your group. Then, **draw** your answer.

Red Peppers Growing

SEP Analyzing and Interpreting Data

12

Do peppers look like this when they are young plants?
Draw a picture of a young pepper plant.

1.1 Plants for a Purpose | 13

1.1 | Wonder How do different plants change over time?

Activity 4
Observe Like a Scientist

Quick Code: us1013s

Flowers Everywhere

Watch the video. **Look** for examples of flowering plants.

Flowers Everywhere

Talk Together

Now, talk together about flowers. Talk about some examples of flowers in your home or at school.

SEP Analyzing and Interpreting Data

14

Activity 5
Analyze Like a Scientist

Quick Code: us1014s

Plants in a Salad

Read about plants in a salad.

 Read Together

Plants in a Salad

You make a salad by taking parts from different **plants**.
What types of plants can you find outside your school?
Are they green?
Are they hard?
Are they soft?
Do they have **flowers**?
How are they alike?
How are they different?

Garden Salad

SEP Developing and Using Models

1.1 Plants for a Purpose | 15

Choose a **fruit** or vegetable used in a salad.

Draw a picture of the plant it comes from.

Activity 6

Evaluate Like a Scientist

Quick Code: us1015s

What Do You Already Know About Plants for a Purpose?

Plant Changes

Read the question. **Circle** all of the correct answers.

The class is planting a butterfly garden. How will the plants change over time? The plants might...

die

grow taller

make seeds

become a seed

get more leaves

move to a new place

1.1 Plants for a Purpose | 17

1.1 | Wonder — How do different plants change over time?

Plant Cycles

A plant is growing in a butterfly garden. **Look** at each image. **Write** 1 in the circle for the image that comes first. **Write** 2, 3, 4 in the circles for the next images. **Write** 5 in the circle for the image that comes last.

Alike and Different

Some plants have the same parts as other plants. **Draw** a line from each plant to all of the words that describe it.

cones

flowers

fruit

leaves

tree

1.1 Plants for a Purpose | 19

1.1 | Learn — How do different plants change over time?

How Can We Group Plants?

Activity 7
Think Like a Scientist

Quick Code: us1016s

Walk in the Park

In this investigation, you will explore the outdoors and record your observations about trees, flowers, and leaves.

What materials do you need?
(per group)

- Pencils
- Paper bags, lunch size
- Hand lens

SEP Planning and Carrying Out Investigations

What Will You Do?

Look at the plants around your school. **Find** two of the same flower. **Draw** and **label** the flowers.

1.1 Plants for a Purpose

1.1 | Learn — How do different plants change over time?

Find three different types of leaves. **Draw** the leaves.

Find two different types of trees. **Write** about the trees.

Pick up three different types of **seeds**. **Draw** the seeds.

Think About the Activity

Compare the flowers, trees, leaves, and seeds you found. **Write** about their size, shape, color, texture, and where they are found.

How They Are The Same	How They Are Different
Flowers	
Leaves	
Trees	
Seeds	

1.1 Plants for a Purpose

1.1 | Learn — How do different plants change over time?

What were some things that other groups found that your group did not?

What patterns did you see?

Activity 8
Observe Like a Scientist

Plant Parts

Quick Code: us1017s

Complete the interactive. Then, **answer** the questions.

What do seeds need to sprout?

_____
_____
_____

SEP Analyzing and Interpreting Data

1.1 Plants for a Purpose | 25

1.1 | Learn — How do different plants change over time?

What do new plants need to grow?

What are five different parts of a plant?

Choose a plant you saw during your walk in the park.
Draw the plant. **Label** the five parts of the plant.

1.1 Plants for a Purpose | 27

1.1 | Learn   How do different plants change over time?

 **Activity 9**
Observe Like a Scientist

Quick Code:
us1018s

Watering Plants

Look at the image. **Compare** the plants in the image with the plants at your school.

Watering Plants

 **Talk Together**

Now, talk together about whether the plants in this picture look like plants at your school.

28

Look at the leaves and seeds you collected during your Walk in the Park. Work together to **sort** the plants in different ways.

Draw one way you grouped the plants. **Label** each group.

SEP Engaging in Argument from Evidence

1.1 Plants for a Purpose | 29

Activity 10
Analyze Like a Scientist

Quick Code: us1019s

Plant Characteristics

Read about plant characteristics.

Read Together

Plant Characteristics

Plants are special because they make their own food. Plant leaves have pores that take in air. The leaves also absorb light from the sun.

Plant roots grow into the soil. Plant roots absorb water and nutrients from the soil. Plants use the power in sunlight to mix air and water and make their own food.

Leaves in Sunlight

SEP Engaging in Argument from Evidence

CCC Structure and Function

Plant roots are important in another way. A plant's roots hold it firmly in the ground so the plant does not blow away.

Roots in Soil

Plant **stems** keep the plant standing up straight. When a plant is tall and straight, more sunlight reaches its leaves. Stems also move water and food to all parts of the plant.

Stem Standing Tall

1.1 Plants for a Purpose

Draw a picture of a young plant. **Label** the stem, a root, and a leaf.

Complete the table by writing about each part of a plant.

Plant Part	Description
leaf	
root	
stem	

1.1 | Learn How do different plants change over time?

How Are Plants the Same and Different?

Activity 11
Observe Like a Scientist

Quick Code: us1020s

Comparing Plants

Watch the videos. Then, **answer** the questions.

Plants in the Desert

How do desert plants compare to plants at your school?

_____
_____
_____

Fruits and Vegetables (Video)

What are some differences between fruit and vegetable plants?

_____
_____
_____
_____
_____
_____

SEP Using Mathematics and Computational Thinking

CCC Patterns

1.1 | Learn How do different plants change over time?

Kinds of Plants (Video)

How are different types of plants the same?

Select two different plants from the videos. **Compare** them using numbers and words.

Complete the table. Choose your own **traits** to compare.

Trait	Plant 1	Plant 2
Type of plant		
Height		

1.1 Plants for a Purpose

1.1 | Learn — How do different plants change over time?

Draw a graph to compare one feature of the two plants.

How are plants the same and different? Use evidence in your answer.

What Plants Can We Eat?

Activity 12
Observe Like a Scientist

Quick Code: us1021s

Fruits and Seeds

Watch the video. **Look** for parts of plants that you can eat.

Fruits and Seeds

Talk Together

Now, talk together about fruits and seeds. Talk about some examples of fruits and seeds that you have eaten.

1.1 Plants for a Purpose

Edible Plants

Plants that can be eaten are called **edibles**. Some plants such as carrots and parsnips have edible roots.

Plants such as celery have edible stems. Spinach and chard have edible leaves.

Broccoli and cauliflower have edible flowers. Apples and bananas are the edible fruits of plants. Nuts and beans are the seeds of plants and also can be eaten.

Fruit: Bottle Gourd

Activity 13
Analyze Like a Scientist

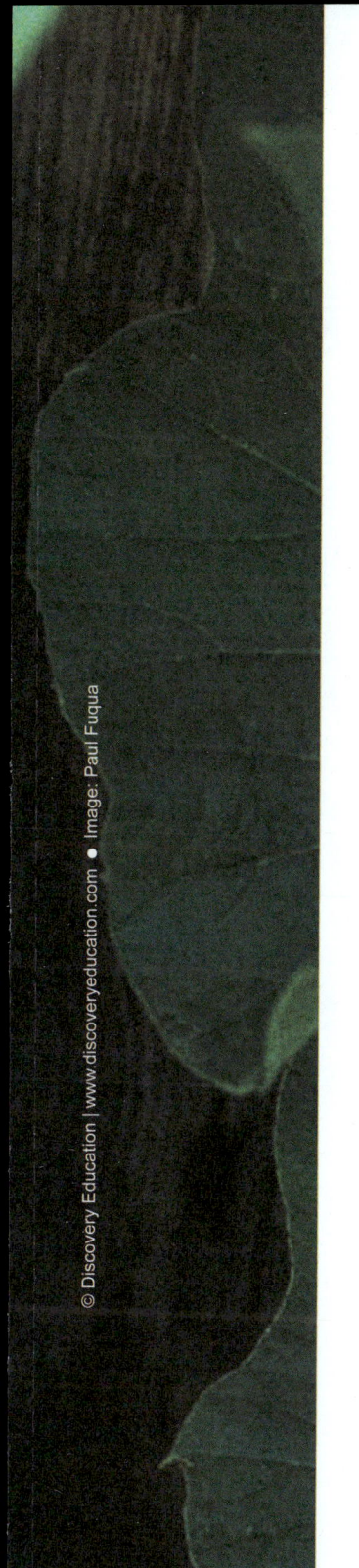

Quick Code: us1022s

Edible Plants

Fill in the T-Chart with examples of edible and non-edible plants.

Edible Plants	Non-edible Plants

1.1 Plants for a Purpose

1.1 | Learn — How do different plants change over time?

Activity 14
Evaluate Like a Scientist

Quick Code: us1023s

Eat Your Plants

Many plants are edible! Imagine that you are going through a restaurant salad bar. **Draw** a picture of a salad that you would eat. Include at least one root, flower, leaf, fruit, and seed in your salad. **Add** labels to show the different parts.

SEP Constructing Explanations and Designing Solutions

How Are Young Plants and Their Parents the Same?

Activity 15
Analyze Like a Scientist

Quick Code: us1024s

Return to Plants in a Salad

Complete the chart. Give your answer to the question. Give evidence-based reasons for your answer.

Is the Plant You Drew in Activity 5 Young or Old?
The plant I drew is an _____ plant.
Evidence-based Reason 1:
Evidence-based Reason 2:

SEP Constructing Explanations and Designing Solutions

1.1 Plants for a Purpose | 43

Read Together

Young and Old Plants

Young plants grow from seeds. Young plants look like their parents and grow similar to their parents.

Sometimes the parent plant can look different from the young plant.

The parent plant may have fruit or flowers.

Young and Old Eggplants

Activity 16

Analyze Like a Scientist

Quick Code: us1025s

Young and Old Plants

Look at the images. **Circle** the sentence that says how the parent plant is different from the young plant.

How are young plants and old plants alike?

How are young plants and old plants different?

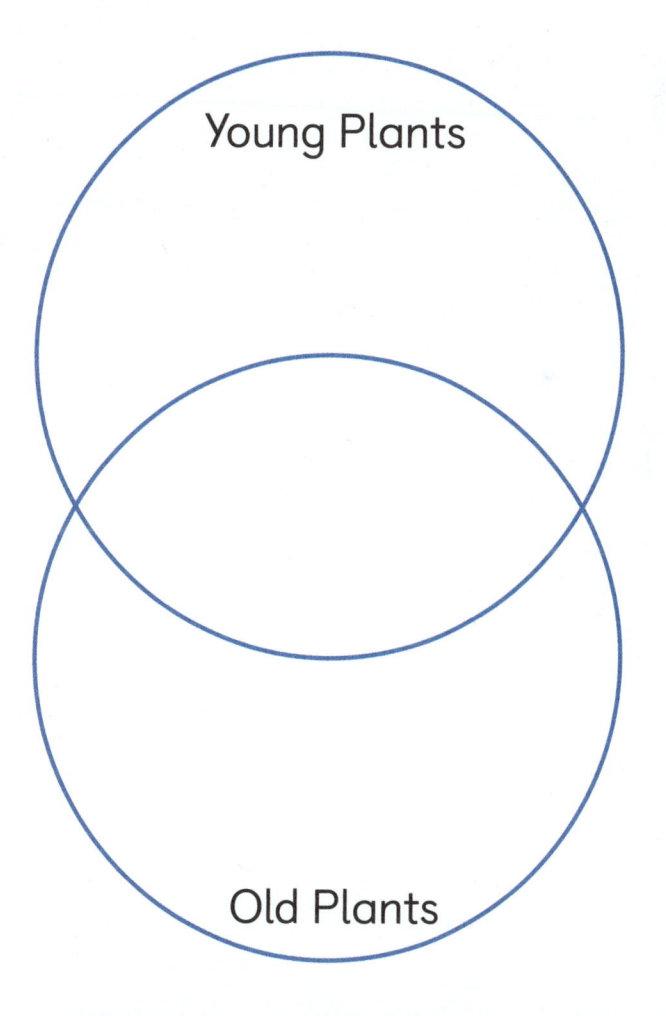

Young Plants

Old Plants

1.1 Plants for a Purpose | 45

1.1 | Learn — How do different plants change over time?

Activity 17
Evaluate Like a Scientist

Quick Code: us1026s

Who Will I Grow Up to Be?

Maya was walking to school and saw two different young plants. She took pictures of the plants. She wondered what each plant would look like when it was older.

Match the correct older plant to each of Maya's pictures.

SEP Constructing Explanations and Designing Solutions

Picture 1 ○   ○ Magnolia Tree

         ○ Fern Plant

Picture 2 ○   ○ Grass

         ○ Pine Tree

1.1 Plants for a Purpose   47

1.1 | **Share** How do different plants change over time?

Activity 18

Record Evidence Like a Scientist

Quick Code: us1027s

Garden Salads

Now that you have learned about the purpose of plants, look again at the picture of the Garden Salad. You first saw this in Wonder.

Let's Investigate Garden Salads

Talk Together

How can you describe Garden Salads now? How is your explanation different from before?

SEP Constructing Explanations and Designing Solutions

48

Look at the Can You Explain? question. You first read this question at the beginning of the lesson.

> **Can You Explain?**
>
> How do different plants change over time?

Now, you will use your new ideas about Garden Salads to answer a question.

1. **Choose** a question. You can use the Can You Explain? question, or one of your own. You can also use one of the questions that you wrote at the beginning of the lesson.

   Your Question

2. Then, **use** the sentence starters to answer the question.

1.1 Plants for a Purpose | 49

1.1 | Share How do different plants change over time?

I know that plants change over time because

Based on my observations, I think

because

STEM in Action

Activity 19
Analyze Like a Scientist

Quick Code: us1028s

Introducing New Foods

Read the text. Then, **answer** the question.

> Read Together

Introducing New Foods

Imagine that you are a farmer. You want to grow a new and improved fruit or vegetable that can solve a problem. Which fruit or vegetable would you choose? What would you change about it? What problem would it solve?

New Fruits

SEP Developing and Using Models

Choose one fruit or vegetable you would like to improve. What would you change and why?

1.1 Plants for a Purpose

What problem will your improvement solve?

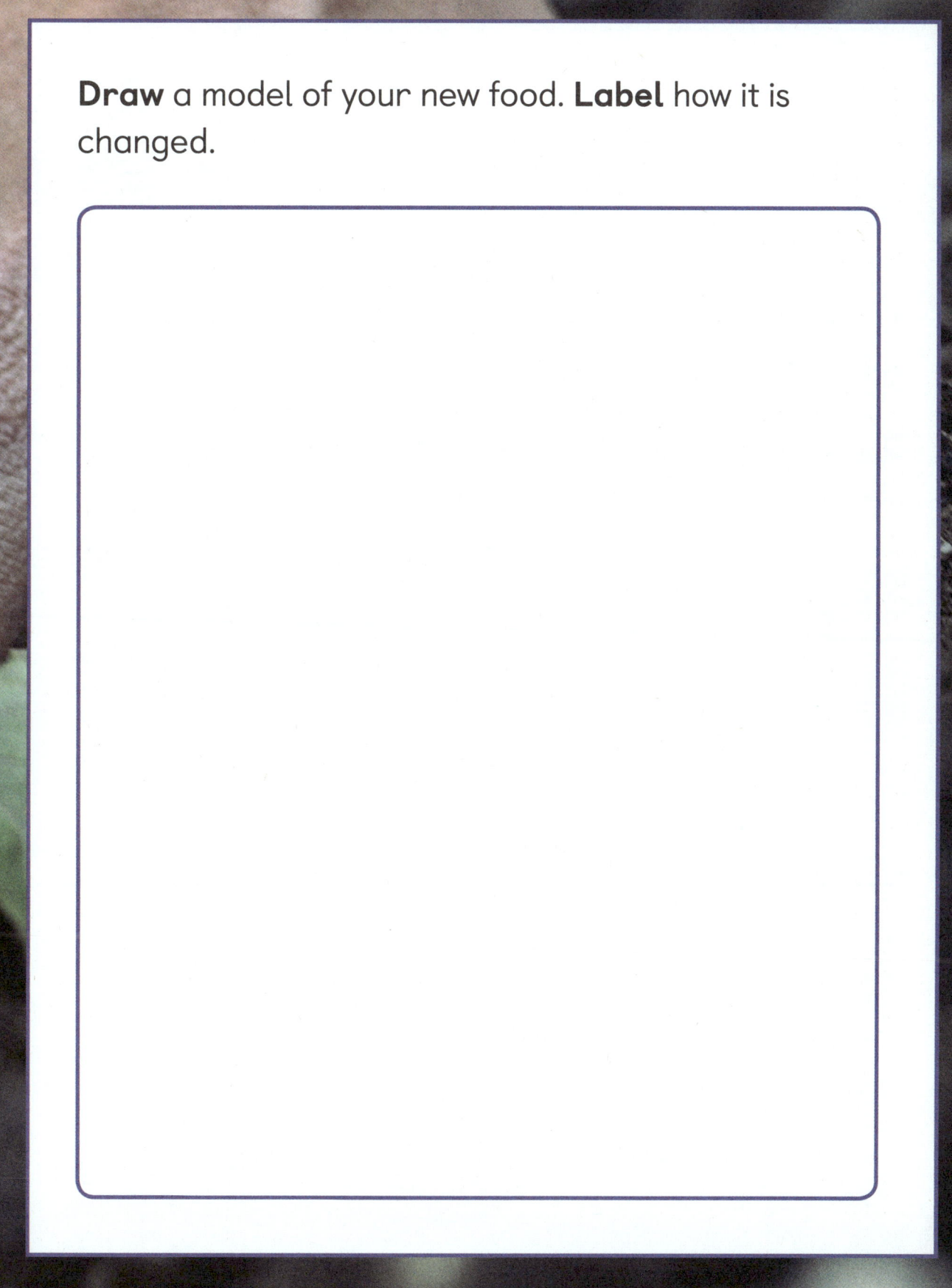

Draw a model of your new food. **Label** how it is changed.

1.1 Plants for a Purpose | 55

1.1 | Share — How do different plants change over time?

Activity 20
Evaluate Like a Scientist

Quick Code: us1029s

Review: Plants for a Purpose

Think about what you have read and seen. What did you learn?

Draw what you have learned. Then, **tell** someone else about what you learned.

Talk Together

Think about what you saw in Get Started. Use your new ideas to discuss how plants in a garden change over time.

1.1 Plants for a Purpose

CONCEPT
1.2

Growing Plants

Student Objectives

By the end of this lesson:

- ☐ I can model how plants use their parts to grow and survive.
- ☐ I can investigate how plants respond to the world around them.
- ☐ I can tell what happens when different objects are put in a light beam.

Key Vocabulary

- ☐ absorb
- ☐ nutrient
- ☐ leaf

Quick Code: us1031s

1.2 Growing Plants

Activity 1
Can You Explain?

How does a plant grow?

Quick Code: us1032s

1.2 Growing Plants   61

1.2 | Wonder — How does a plant grow?

Activity 2
Ask Questions Like a Scientist

Quick Code: us1033s

Hydroponic Plants

Look at the photo. **Think** of questions you have about the picture.

Let's Investigate Hydroponic Plants

SEP Asking Questions and Defining Problems

62

Write or **draw** in the chart to show what you know, wonder, and learned about plants as you complete the lesson.

Topic: Hydroponic Plants

I <u>K</u>now	I <u>W</u>onder	I <u>L</u>earned

1.2 Growing Plants

1.2 | Wonder — How does a plant grow?

Activity 3
Think Like a Scientist

Quick Code: us1034s

Inspecting Your Spot

In this activity, you will use an observation area to study the environment.

What materials do you need?
(per group)

- Paper bags, lunch size
- Yarn
- Soap for washing hands after

SEP Planning and Carrying Out Investigations
CCC Patterns

What Will You Do?

You are going to observe a small area outside on the ground. What will you find in your area? **Write** or **draw** your predictions.

Draw a picture of your area. **Collect** samples of things that you see.

1.2 Growing Plants | 65

1.2 | Wonder How does a plant grow?

Draw a picture of any animals you find in your area.

Draw a picture of the plants you find in your area.

Tell about any nonliving objects that are in your area.

Think About the Activity

Compare the things you saw in your area. How are plant and animal needs the same? How are plant and animal needs different?

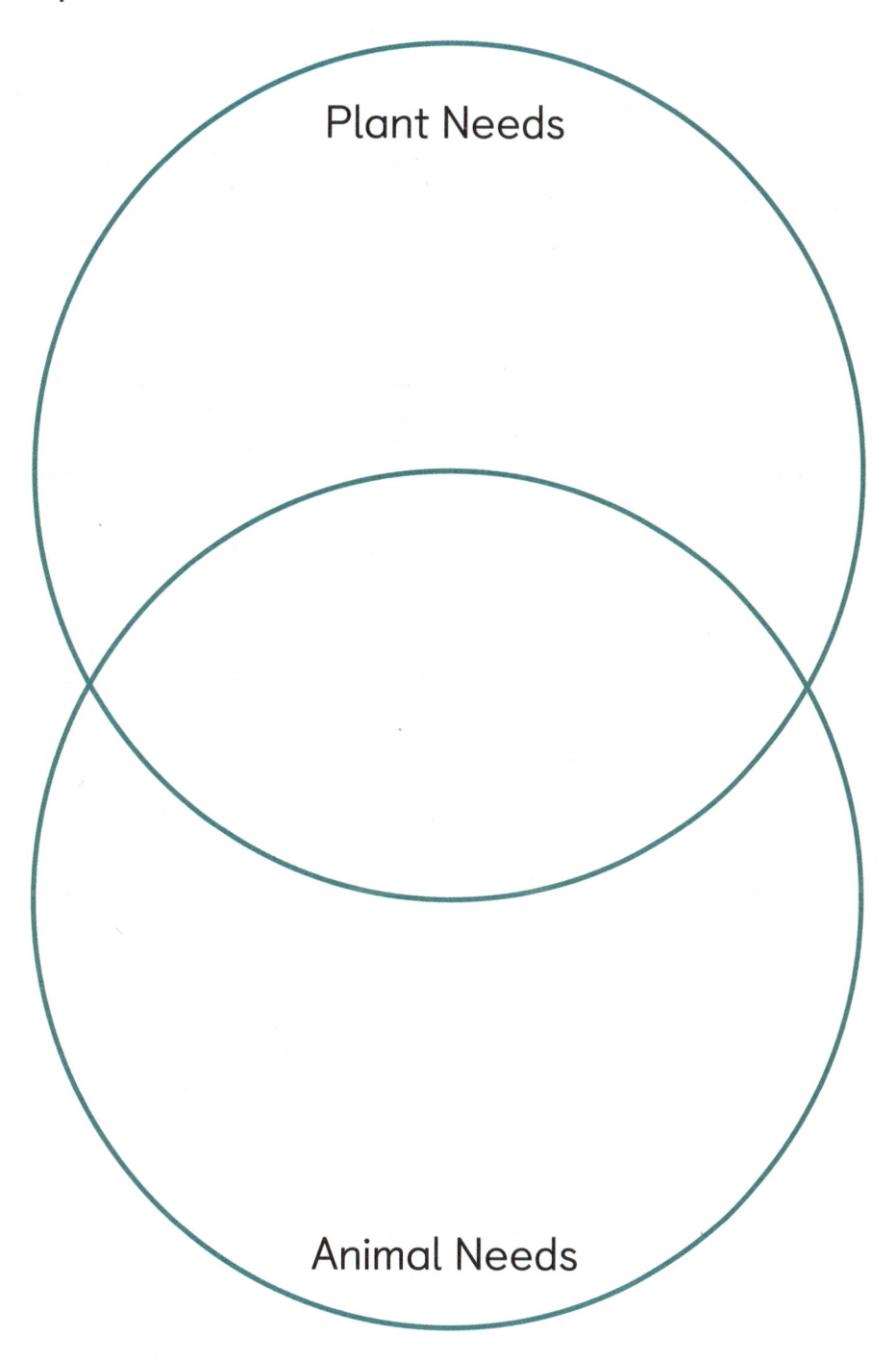

Plant Needs

Animal Needs

1.2 Growing Plants | 67

1.2 | Wonder — How does a plant grow?

Why is it important to use your senses in your area?

Why is it important for scientists to ask questions?

Activity 4
Observe Like a Scientist

Quick Code: us1035s

Plant Needs

Watch the video. **Look** for ways plants and living things are the same or different. Add any new information to your chart for Activity 2: Hydroponic Plants.

Plants and Other Living Things

Talk Together

Now, talk together about living things. How are plants the same as other living things? How are they different?

1.2 Growing Plants

1.2 | Wonder How does a plant grow?

Watch the video. **Look** for the needs of living things.

Basic Needs

💬 Talk Together

Now, talk together about the needs of living things. How do you meet your basic needs? How does this compare to plants?

Write what you learned in the chart in Activity 2: Hydroponic Plants.

Activity 5
Evaluate Like a Scientist

Quick Code: us1036s

What Do You Already Know About Growing Plants?

Which Part?

Each part of a plant has a job. **Label** each part of the drawing using the letters for each job.

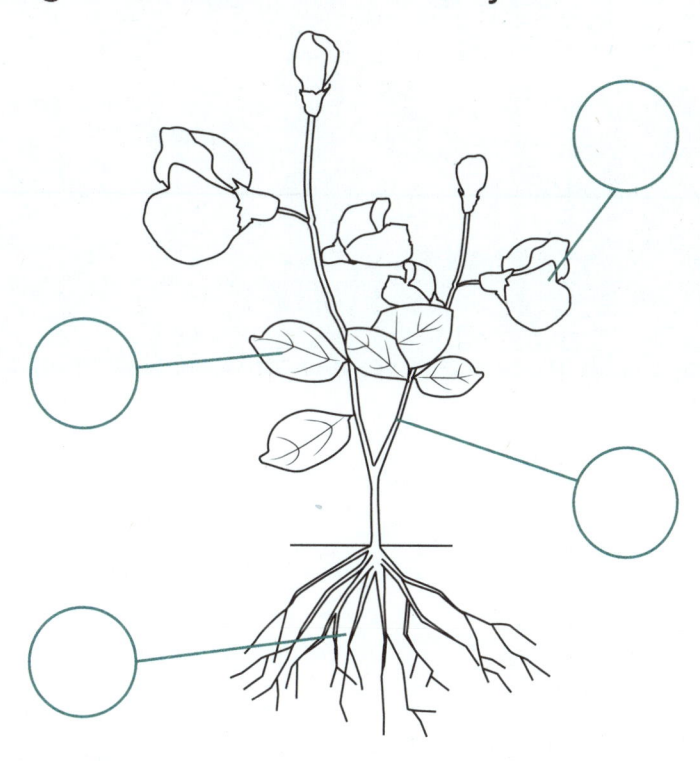

A. carries water to other plant parts

B. takes in water from the soil

C. makes seeds so new plants can grow

D. takes in sunlight so the plant can make food

SEP Engaging in Argument from Evidence

CCC Structure and Function

1.2 Growing Plants | 71

1.2 | **Wonder**   How does a plant grow?

What Do Plants Need?

What do plants need to live and grow? **Look** at each image. **Circle** all of the images that show something plants need.

1.2 Growing Plants | 73

1.2 | Learn — How does a plant grow?

What Do Plants Need to Grow?

Activity 6

Investigate Like a Scientist

Quick Code: us1037s

Hands-On Investigation: Growing Seeds

In this activity, you will work in small groups. Each group will try a different way to grow seeds.

Make a Prediction

Do you think your seeds will sprout? **Write** or **draw** your prediction.

SEP Planning and Carrying Out Investigations
CCC Cause and Effect

What materials do you need?
(per group)

- Seeds, kidney beans
- Paper towels
- Plastic bags gallon size, zip-closure
- Water

What Will You Do?

Plant your seeds following your teacher's instructions.

Draw a picture of what you did.

1.2 | Learn How does a plant grow?

Observe your seeds every day for a week. **Fill** in one line of the chart for each day.

Date	Drawing	Observation

Compare your data table with the other groups in the class.

What patterns do you observe in how seeds grow?

Which seeds grew best? Why do you think they grew best?

1.2 Growing Plants

1.2 | Learn How does a plant grow?

Now, your group will design your own experiment on how seeds grow. First, **watch** the video.

Video: Growing a Bush Bean

Now, think of a way you can change only one thing to see how it affects the seeds. What will you change in your experiment? How will you change it?

Observe your experiment every day for a week. **Fill** in one line of the chart for each day. Label the different conditions your seeds are in.

Date	Drawing	Observation

1.2 Growing Plants

1.2 | Learn — How does a plant grow?

Think About the Activity

Fill in the T-Chart with what you learned in your experiments. Be sure to give evidence for each claim.

What do seeds need to grow?

What Seeds Need to Grow	Evidence

What did you learn from the experiment your group designed?

What else would you like to learn about sprouting seeds?

1.2 | Learn How does a plant grow?

Activity 7
Observe Like a Scientist

Quick Code: us1038s

Growing Plant

Watch the video without sound. **Look** for how a plant grows from a seed.

Video

Growing Plant

Talk Together

Now, talk together about how the seed grew in the video. Tell what you observed.

Activity 8
Analyze Like a Scientist

Quick Code: us1039s

What Do Plants Need?

Read about plants' needs. Then **complete**, the activity that follows.

> Read Together

What Do Plants Need?

What makes a plant grow?

Plants need **nutrients**. In soil and sunlight to grow. Most plants get water and nutrients from the soil through their roots.

The roots also keep the plant in one place and hold up the stem. The stem carries the nutrients from the roots to the **leaves**.

Seedling with Roots

SEP Obtaining, Evaluating, and Communicating Information

1.2 Growing Plants | 83

Read Together

Birch Leaves

Bee and Flower

Sunlight hits the leaves and helps the leaves make food for the plant. The leaves use sunlight, water, and air to make their own food. This food helps the plant grow tall and strong.

Some plants grow flowers to attract bees. Bees help flowers make seeds. The seeds will grow into new plants.

Draw a picture of a young plant. **Label** the parts of the plant. **Write** how the plant meets its needs.

1.2 Growing Plants

1.2 | Learn How does a plant grow?

How Do Parts of a Plant Help the Plant Grow?

Activity 9
Observe Like a Scientist

Quick Code: us1040s

Plant Parts

Complete the interactive. Then, **answer** the questions.

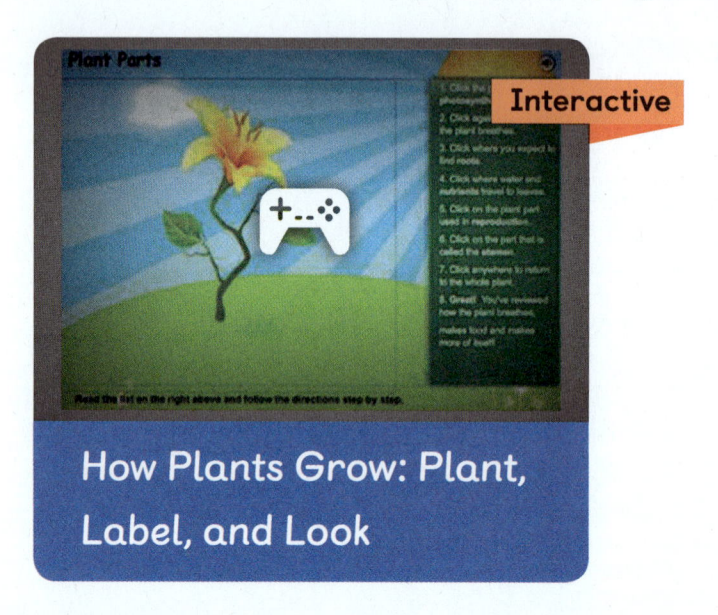

Interactive

How Plants Grow: Plant, Label, and Look

What two functions did leaves have in the activity?

_____
_____
_____
_____

Draw a line to match the plant part with its function.

flower	sends water and nutrients to leaves
root	takes in water and nutrients
stem	used in reproduction

Draw a flower. **Label** a stamen in the flower.

CCC **Structure and Function**

1.2 Growing Plants | 87

Activity 10
Analyze Like a Scientist

Quick Code: us1041s

Parts of a Plant

Choose a word.

| seed | soil | water | air | sun | plant |
| roots | nutrients | stem | leaves |

Read about the parts of plants. When you hear the word you chose, **act** it out.

Read Together

Parts of a Plant

Sprouting Seed

Plants begin to grow when a seed is buried in soil.

With water, air, and light, the seed will grow into a plant.

Each part of a plant helps the plant grow.

Roots hold the plant in the ground and help it get water and nutrients.

The stem holds the plant up.

The leaves of the plant collect sunlight to make food.

Write or **draw** about how plant parts help plants stay alive.

Structure	Function

SEP Obtaining, Evaluating, and Communicating Information

CCC Structure and Function

1.2 Growing Plants

1.2 | Learn — How does a plant grow?

Activity 11
Evaluate Like a Scientist

Quick Code: us1042s

Marcus's Plant Care Business

Read the problem about Marcus's plant care business. Then **complete** the activity.

Marcus is starting a business to take care of his neighbors' plants when they go on vacation. When he sees a plant with a problem, he tries to figure out which part is hurt. Help Marcus remember the jobs plant parts do to help the plant grow.

CCC Structure and Function

Look at the list of plant problems. **Write** an X in the boxes for each part that might be hurt and causing the problem. You may write more than one X for each problem.

Plant Part	Plant Does Not Get Water From Soil	Plant Does Not Make Food	Plant Will Not Make Seeds	Plant Fell Over
flower				
leaves				
roots				
stem				

1.2 Growing Plants | 91

1.2 | Learn How does a plant grow?

How Does Sunlight Reach Plants?

Activity 12
Investigate Like a Scientist

Quick Code: us1043s

Hands-On Investigation: Shady Plants

In this activity, you will choose different types of leaves. You will draw their shadows.

Make a Prediction

You are going to compare the shadows of two different leaves. **Look** at the two leaves you will use. **Write** or **draw** your predictions.

Which leaf will make the darkest shadow? Why?

SEP Planning and Carrying Out Investigations

CCC Structure and Function

What materials do you need?
(per group)

- Plastic bags gallon size, zip-closure
- Various opaque leaves (ex: hazelnut)
- Various translucent leaves (ex: maple)
- White paper
- Pencils
- Flashlight
- Batteries, size D

What Will You Do?

Place one opaque leaf and one translucent leaf in the bag. **Lie** both leaves flat. **Draw** a picture of the two leaves.

1.2 Growing Plants | 93

1.2 | Learn How does a plant grow?

Hold the bag between the light and a piece of paper. **Trace** the whole shadow. **Shade** the darker part of each shadow.

What causes a shadow?

Why are some parts of the shadow darker than others?

Which leaf let more light through? How do you know?

Think About the Activity

Compare the shadows of the opaque and translucent leaves.

How are shadows of the translucent and opaque leaves the same? How are shadows of the translucent and opaque leaves different?

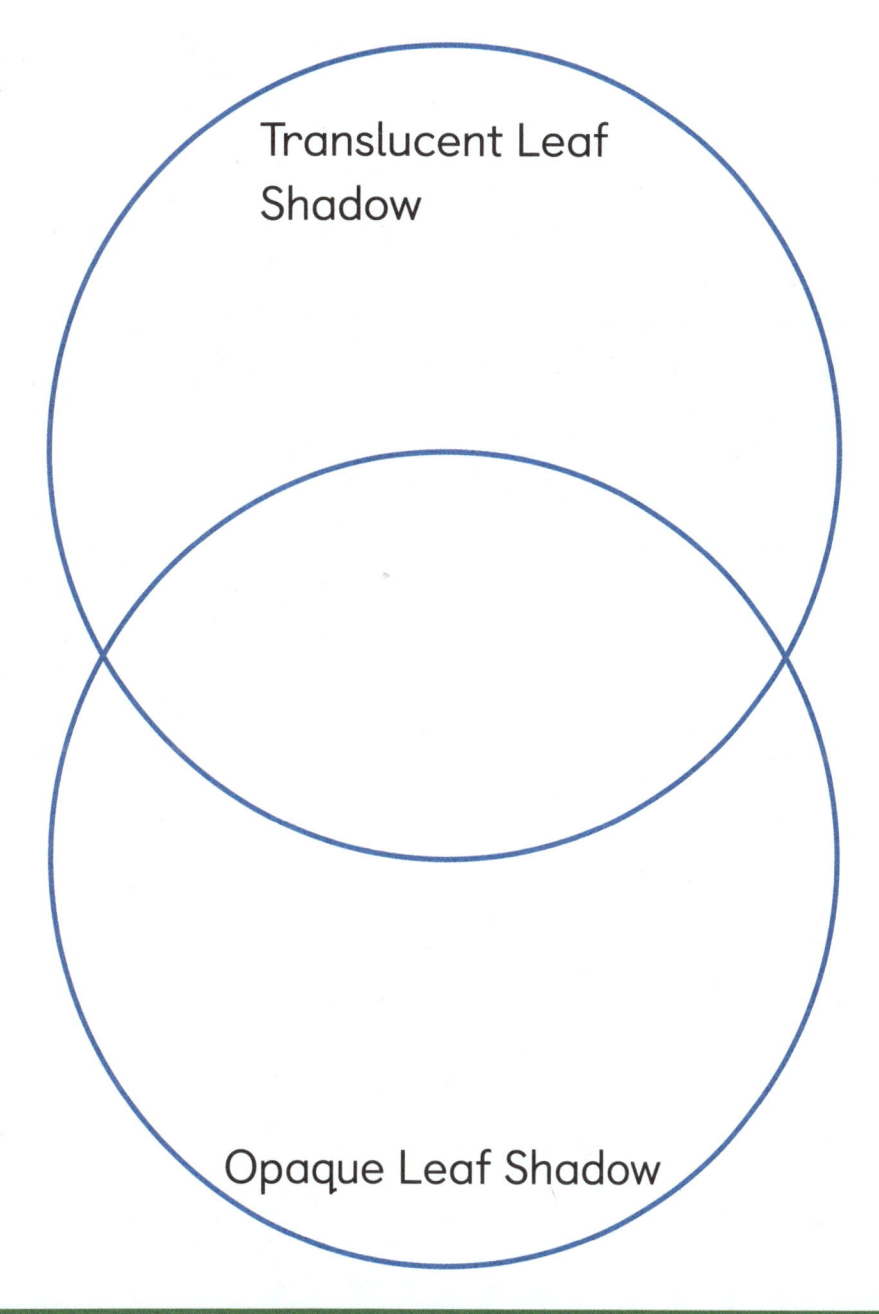

Translucent Leaf Shadow

Opaque Leaf Shadow

1.2 Growing Plants

1.2 | Learn How does a plant grow?

How could you use what you learned in this activity in the real world?

How might the amount of light that goes through a leaf affect other plants?

Activity 13

Observe Like a Scientist

Quick Code: us1044s

Plants Under Cover

Look at the image. **Talk** about which plants get more sunlight and less sunlight. **Talk** about which plants get more water and less water. **Draw** a line from each label to a plant in the picture.

Plants Under Cover

| more sunlight | less sunlight | more water | less water |

1.2 Growing Plants | 97

1.2 | Learn How does a plant grow?

How do plants found under other plants get the sunlight they need?

How does Activity 12: Shady Plants support your answer?

- **SEP** Engaging in Argument from Evidence
- **SEP** Constructing Explanations and Designing Solutions
- **CCC** Structure and Function

Activity 14
Evaluate Like a Scientist

Quick Code: us1045s

The Most Light

Circle the plant in the image that is getting the most light.

SEP Analyzing and Interpreting Data

1.2 Growing Plants | 99

1.2 | Share How does a plant grow?

Activity 15
Record Evidence Like a Scientist

Quick Code: us1046s

Hydroponic Plants

Now that you have learned about growing plants, look again at the picture of the Hydroponic Plants. You first saw this in Wonder.

Let's Investigate Hydroponic Plants

Talk Together

How can you describe Hydroponic Plants now? How is your explanation different from before?

SEP Constructing Explanations and Designing Solutions

Look at the Can You Explain? question. You first read this question at the beginning of the lesson.

> 💬 **Can You Explain?**
> How does a plant grow?

Now, you will use your new ideas about Hydroponic Plants to answer a question.

1. **Choose** a question. You can use the Can You Explain? question, or one of your own. You can also use one of the questions that you wrote at the beginning of the lesson.

Your Questions

2. Then, **use** the sentence starters to answer the question.

1.2 Growing Plants | 101

1.2 | Share How does a plant grow?

Plants grow by

Based on my observations, I think

because

An example of a part of a plant that helps it grow would be

The evidence I collected shows

STEM in Action

Quick Code: us1047s

Activity 16
Analyze Like a Scientist

Growing Plants without Soil

Read the story. Then, **complete** the activity.

Read Together

Growing Plants without Soil

Most plants grow in soil. They get water from the soil. Soil holds them in place. But some plants are not grown in soil. Some are grown in water. This is called *hydroponics*. Plants are placed in trays. The trays are filled with water. The water has nutrients in it. Nutrients are what plants need to grow.

Hydroponic Plants

SEP Obtaining, Evaluating, and Communicating Information

Often, these plants grow indoors. People use special lights to help them grow. The lights stay on 24 hours a day, so the plants can grow very fast. They can grow twice as fast as plants outdoors.

Other plants are grown in air. They are placed in special boxes. Their roots may grow into foam. The foam contains nutrients. The roots may also just hang in the air. Plants can be sprayed with mist. The mist gives the plants water. The mist also contains nutrients. These plants are grown indoors. People use special lights on them, too. These plants do not use much water. Saving water helps farmers and it costs less money. The plants will grow faster, too. They do not take up much space. They can grow stacked on one another. This means they can grow in small spaces.

Read Together

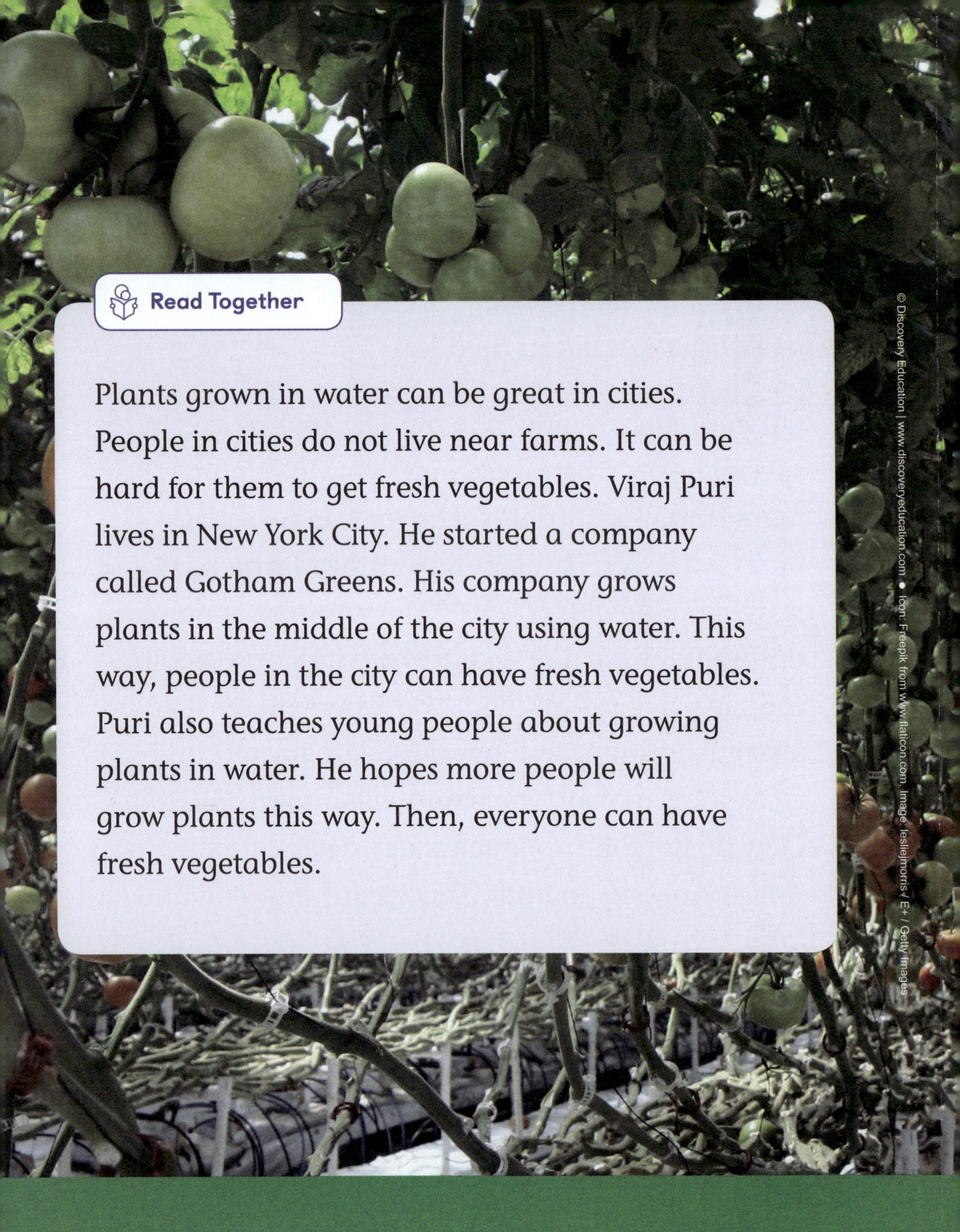

Plants grown in water can be great in cities. People in cities do not live near farms. It can be hard for them to get fresh vegetables. Viraj Puri lives in New York City. He started a company called Gotham Greens. His company grows plants in the middle of the city using water. This way, people in the city can have fresh vegetables. Puri also teaches young people about growing plants in water. He hopes more people will grow plants this way. Then, everyone can have fresh vegetables.

How Is It Grown?

Match the methods of growing plants with the descriptions.

Method	Description
	It uses less water.
	Plant roots grow in air.
Aeroponic	
	Plant roots grow in soil.
Hydroponic	Plants grow very quickly.
	Plant roots grow in water.
In soil	
	It doesn't take up much room.
	Plants get the water and nutrients they need.

1.2 Growing Plants | 107

1.2 | Share How does a plant grow?

Activity 17
Evaluate Like a Scientist

Quick Code: us1048s

Review: Growing Plants

Think about what you have read and seen. What did you learn?

Draw what you have learned. Then, tell someone else about what you learned.

Talk Together

Think about what you saw in Get Started. Use your new ideas to discuss how you can design a garden.

SEP Obtaining, Evaluating, and Communicating Information

1.2 Growing Plants

CONCEPT
1.3

Designing for Plants

Student Objectives

By the end of this lesson:

- ☐ I can use what I know about how plants use their parts to design a solution.
- ☐ I can design a solution to solve a human problem.
- ☐ I can tell if a design solution to a problem will work.

Key Vocabulary

- ☐ light
- ☐ material
- ☐ soil
- ☐ structure
- ☐ system
- ☐ tendril
- ☐ water

Quick Code: us1050s

1.3 Designing for Plants | 111

Activity 1

Can You Explain?

How can humans help plants grow?

Quick Code: us1051s

1.3 Designing for Plants | 113

1.3 | Wonder — How can humans help plants grow?

Activity 2
Ask Questions Like a Scientist

Quick Code: us1052s

Creepers

Look at the photo. **Think** about the **structure** of the creeper plant.

Let's Investigate Creepers

SEP Asking Questions and Defining Problems
CCC Structure and Function

114

Talk Together

Now, talk together about how the creeper's structure is related to its function.

Fill in the T-Chart about the structure and function of the creeper plant. Then, **write** other examples of structure and function pairs in the chart.

Structure of Plants	Functions of Plants

1.3 Designing for Plants

1.3 | Wonder — How can humans help plants grow?

Look at the photo. **Think** about the structure and function of the Venus flytrap.

Venus Flytraps

💬 Talk Together

Now, talk together about how the Venus flytrap's structure is related to its function.

Think about the structure and function of objects in your classroom or in your life. **Write** "I wonder" questions about the objects.

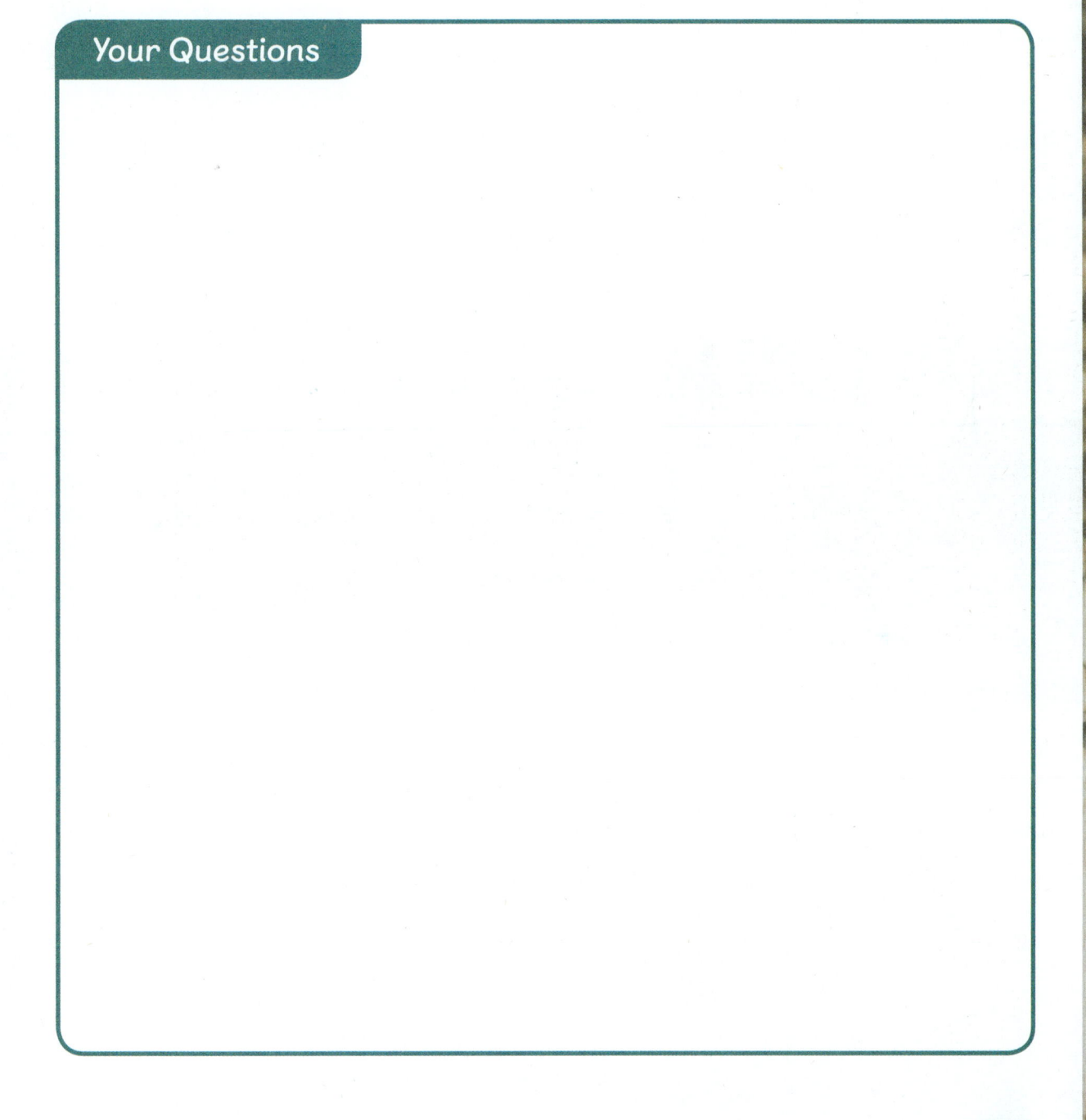

Your Questions

Activity 3
Analyze Like a Scientist

Quick Code: us1053s

Biomimicry

Read about biomimicry. Then, **complete** the activity.

Read Together

Biomimicry

Banyan Fig Tree

Roof Shaped Like a Leaf

Biomimicry happens when designs from nature are used to make things that help us.

The banyan fig tree leaves have pointy tips that help **water** slide off.

The rooftop is designed like Banyan leaves.

The shape of its parts helps water slide off the roof.

The water carries the dirt away.

Beetle

Collecting Water

The bumps on this beetle's back help it to get water from the air.

The water slides to the beetle's mouth.

The bumpy material is useful in places where there is little water.

It helps people collect water from the air, so they can use the water.

1.3 Designing for Plants | 119

Discuss what you read on biomimicry. **Find** an object in your school that copies a design from nature.

Draw a picture of the object. Then, **draw** a picture of the object in nature that it copies.

Activity 4
Evaluate Like a Scientist

Quick Code: us1054s

What Do You Already Know About Designing for Plants?

How Parts Work Together

All of the parts of a tent work together to make a good structure for camping.

Match the image of the material or part to its function.

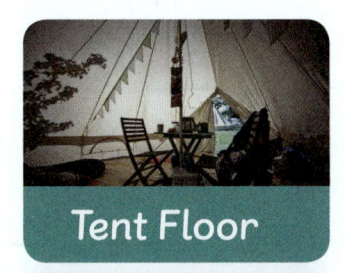

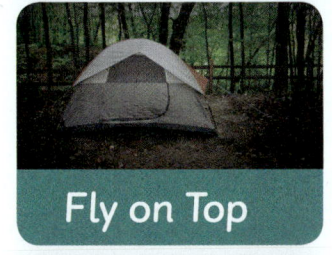

- Tent Floor
- Fly on Top
- Netting
- Zipper

- allows air to flow
- keeps animals out
- protects from ground
- protects from rain

1.3 Designing for Plants

1.3 | **Wonder** How can humans help plants grow?

Breaking Things Down

There are many things in our lives that have been designed to solve a problem. **Think** of two objects in your life. What problems do they solve? Why are they designed like they are? **Write** the objects and the answers.

Object 1:

Object 2:

The Model for Design

Engineers use many words to describe their process to make something new. Some of the terms engineers might use are: create, ask, imagine, plan, improve. Think about what the words mean. Can you put the terms of the process in order from start to finish?

Write the numbers from 1 to 5 to order the steps from start to finish.

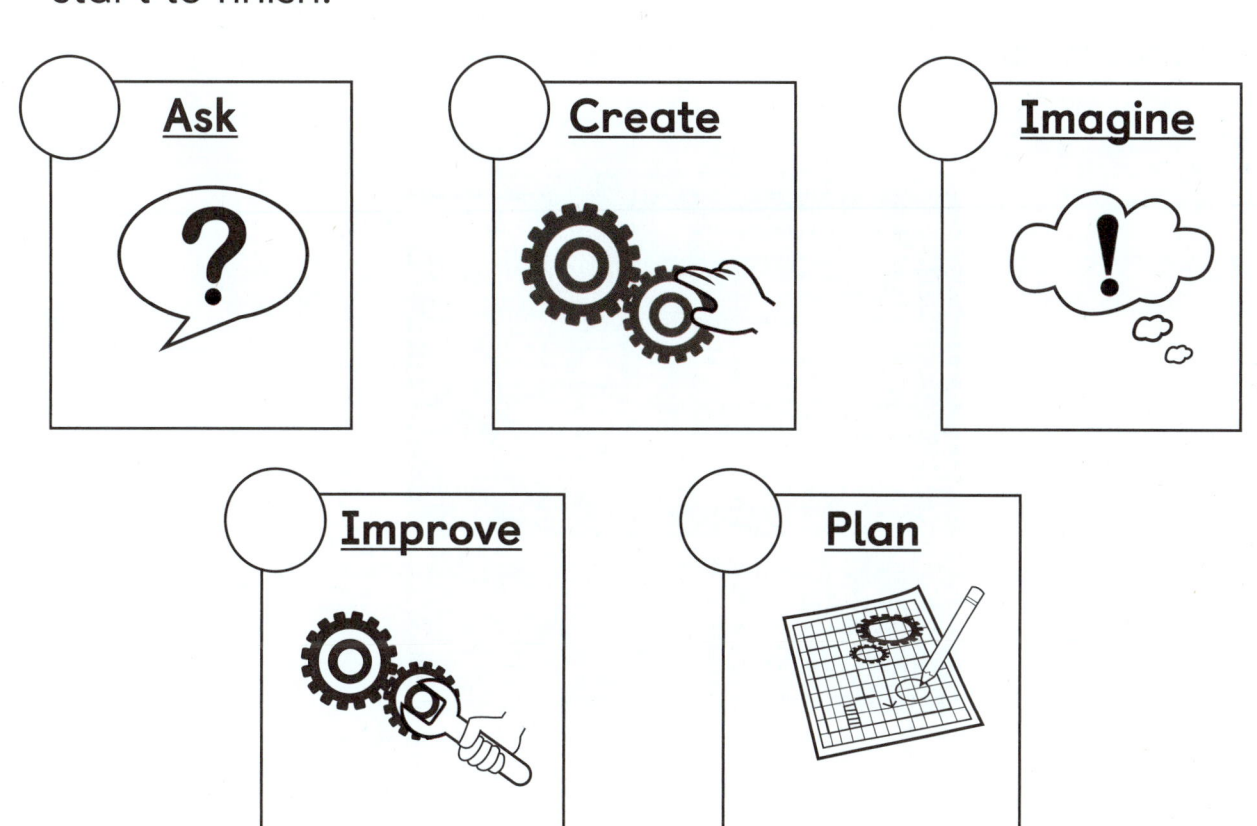

Ask

Create

Imagine

Improve

Plan

1.3 Designing for Plants

1.3 | Learn How can humans help plants grow?

What Can People Build to Help Plants Live and Grow?

Activity 5
Observe Like a Scientist

Quick Code: us1055s

How Does Your Garden Grow? Part 1

Complete the Introduction tab of the interactive.

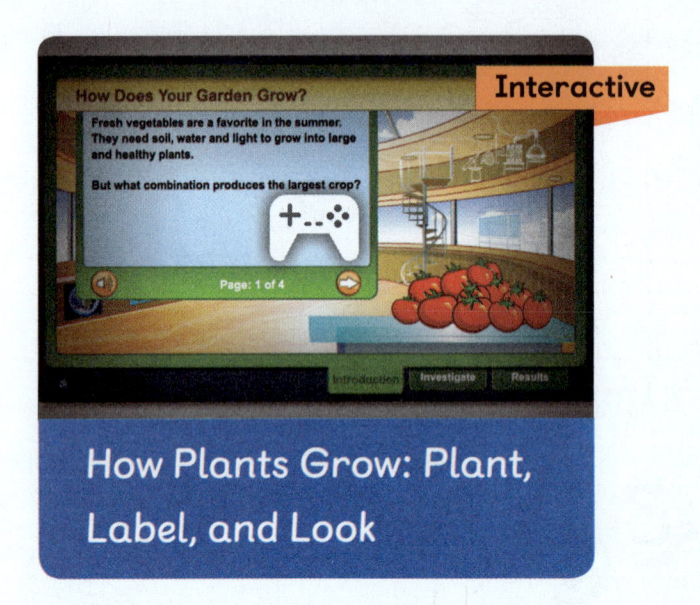

How Plants Grow: Plant, Label, and Look

SEP Planning and Carrying Out Investigations
CCC Cause and Effect

Make a plan for your investigation.

What are the variables, or the things that can change, to grow the tomatoes?

What variable will you change in your investigation?

What question will your investigation test?

1.3 | Learn   How can humans help plants grow?

Make a prediction.
What will happen and why do you think so?

Design a procedure for your investigation.
Write the steps you will take to answer your question.

Activity 6
Observe Like a Scientist

Quick Code: us1056s

How Does Your Garden Grow? Part 2

Complete the investigation you designed in Part 1.
Record your results in the table.

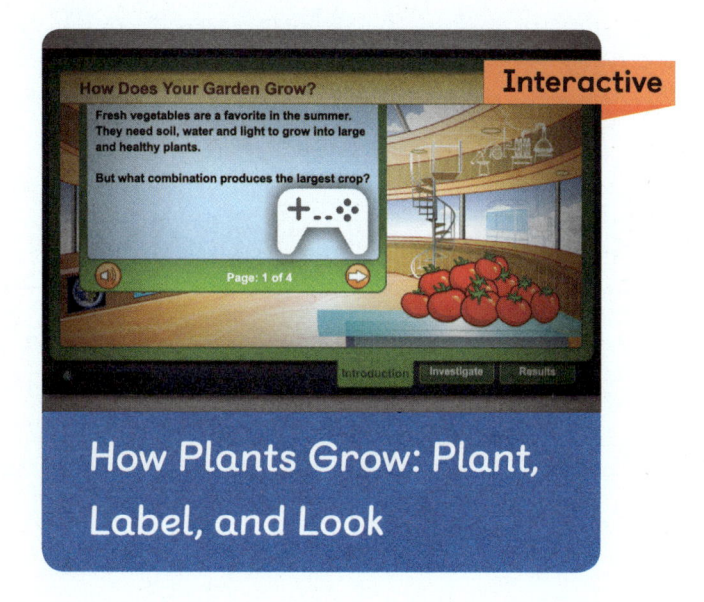

How Plants Grow: Plant, Label, and Look

SEP Planning and Carrying Out Investigations
CCC Cause and Effect

1.3 Designing for Plants | 127

1.3 | Learn   How can humans help plants grow?

Trial	Soil	Water	Light	Tomatoes per Plant	Tomato Size
1					
2					
3					
4					
5					

What happened in the experiment? Did it match your prediction?

1.3 | Learn How can humans help plants grow?

Activity 7
Observe Like a Scientist

Quick Code: us1057s

Gardens

Look at the images of the gardens. **Discuss** how the designs are alike and different. **Think** about how the designs help the plants in the garden grow.

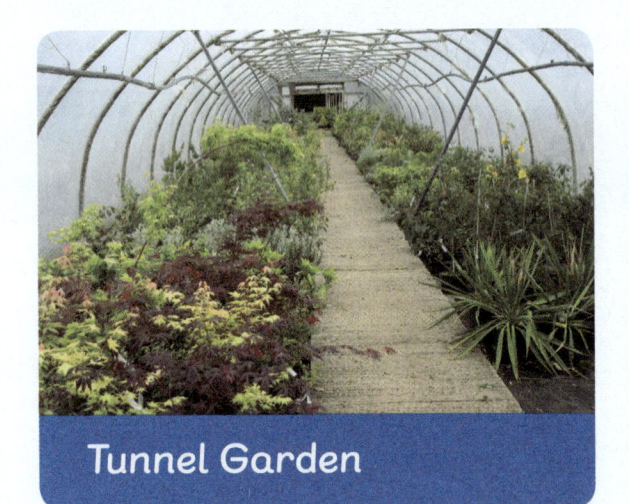

Tunnel Garden

Raised Bed Garden

Greenhouse

Living Wall

CCC Structure and Function

Choose two of the images. **Write** the title of the images in each circle.

How are the designs of each garden the same?
How are the designs of each garden different?

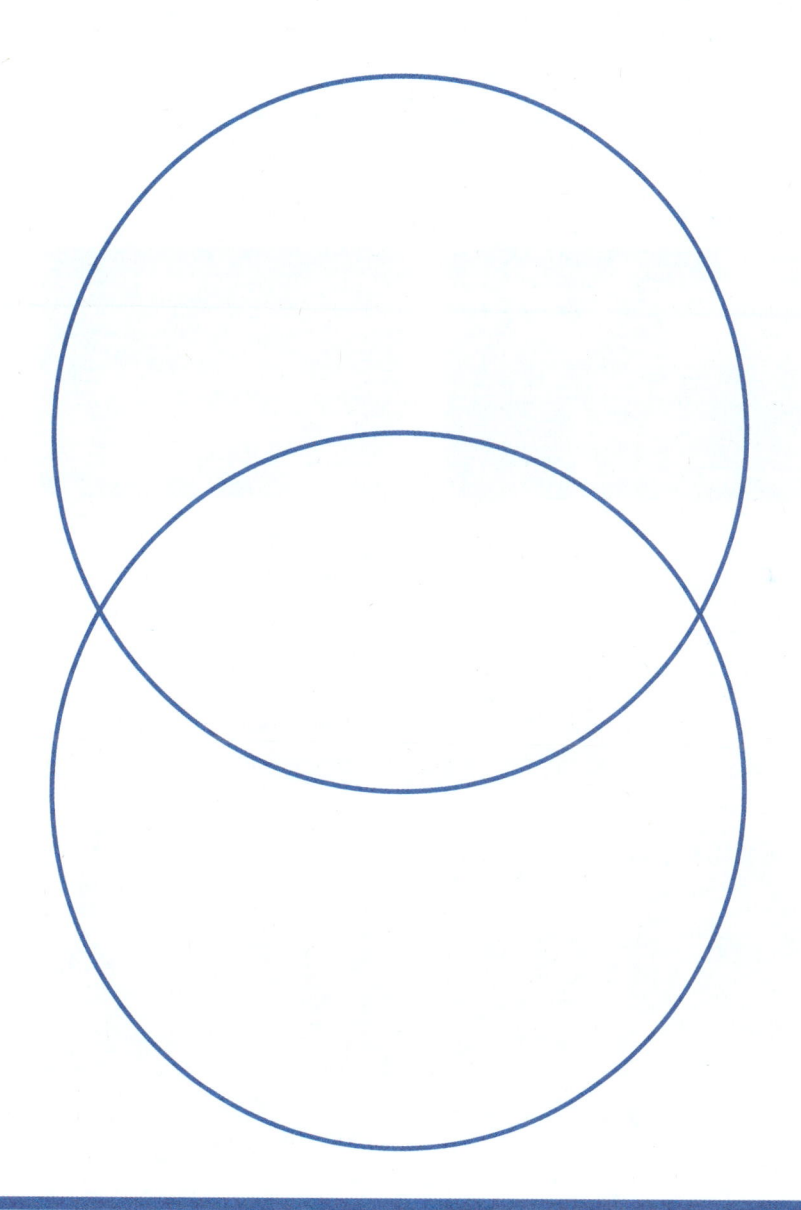

1.3 Designing for Plants

1.3 | Learn   How can humans help plants grow?

Activity 8
Observe Like a Scientist

Quick Code: us1058s

Planting a Garden

The gardener wants to plant a new vegetable garden. He wants to put it in a place where his plants will have all their needs met. Where should he place his garden?

Circle the image that shows the best place to locate the garden.

SEP Constructing Explanations and Designing Solutions

SEP Engaging in Argument from Evidence

What Can People Learn from Plants to Help Plan and Create Structures?

Activity 9
Think Like a Scientist

Quick Code: us1059s

Shapes of Plants

In this activity, you will observe and draw plants in your schoolyard. You will then figure out the shapes of the plants.

What materials do you need?
(per group)

- Metric ruler
- Paper
- Pencils
- Cut-out shapes

SEP Developing and Using Models
CCC Structure and Function

1.3 Designing for Plants | 133

1.3 | Learn How can humans help plants grow?

What Will You Do?

Review the parts of plants with your teacher. **List** the main parts of plants and their functions.

Your teacher will assign you and your partner a plant. **Draw** your plant.

Match shapes to different parts of your drawing.
Write the shapes you used.

1.3 Designing for Plants

1.3 | Learn   How can humans help plants grow?

Think About the Activity

Compare your drawing with your partner's drawing. **Look** at other drawings of other plants. **Discuss** patterns you see in the drawings.

Fill in the T-Chart with what you learned from the drawings. **Match** the shapes and structures in the drawing with the function.

Shapes I Found in Plants	How the Shapes Helped the Plant

How was your drawing different from the real plant that you drew?

What patterns did you notice in your group and other groups' drawings?

Activity 10
Analyze Like a Scientist

Quick Code: us1060s

Designing Structures

Read about designing structures. Then, **answer** the questions.

> **Read Together**

Designing Structures

Architects and engineers use nature to help them design structures.

Trees have a large trunk section with branches that reach out in all directions. Engineers and architects have used the shape of a tree to design structures that hold up large surfaces.

Fern Tree from Below

Airport

138

Structures include more than one part.

Like parts of plants, each part of a structure has a job to do.

In a greenhouse, the roof absorbs and lets in **light** to the plants.

Like roots that take water to the stem of a plant, gardeners design **systems** to move water to the plants in the greenhouse.

These systems are called irrigation systems.

Mangrove Roots

Train Station

1.3 Designing for Plants

Choose the fern image or the mangrove image. What everyday item is like the structure of the plant?

Choose the airport image or the train station image. How is the structure like a part of a plant?

CCC **Structure and Function**

Activity 11
Evaluate Like a Scientist

Quick Code: us1061s

Which Part Fits Where?

Match the image to the structure in which it is a part.

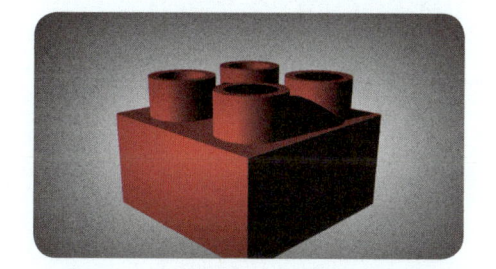

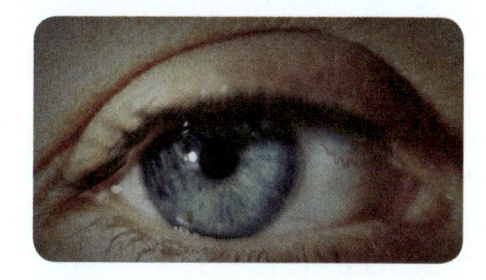

CCC Structure and Function

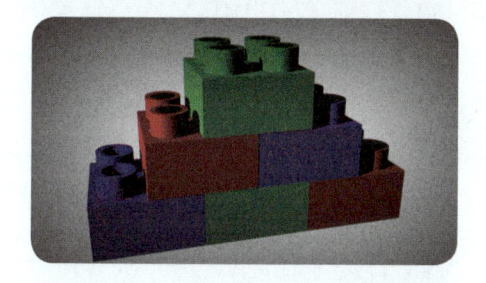

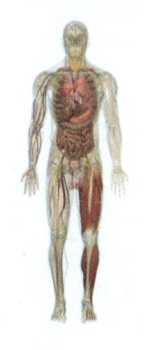

1.3 Designing for Plants | 141

1.3 | Learn — How can humans help plants grow?

How Do People Choose What They'll Use?

Activity 12
Investigate Like a Scientist

Quick Code: us1062s

Hands-On Investigation: The Right Stuff

In this activity, you will explore the properties of different **materials** and use the materials to make different things.

Make a Prediction

Write or **draw** your prediction. What materials will be the most useful for giving an object a stable shape?

SEP Constructing Explanations and Designing Solutions
CCC Structure and Function

What materials do you need?
(per group)

- Felt rectangles with matching holes punched along the side, 2
- Metal brads, 6
- Aluminum foil
- Cardboard paper towel tube
- Cotton balls
- Craft sticks
- Glue sticks
- Tissue paper
- Construction paper
- Markers
- Yarn
- Clear tape
- Scissors
- Metric ruler
- Connecting cubes
- Empty plastic soda/water bottles from home (washed)
- Empty soda cans from home (washed)

What materials will be the most useful for holding materials together?

1.3 | Learn How can humans help plants grow?

Review the garden images. **Observe** the parts of each garden system. **Think** about the function of each part. What function does each part serve?

Greenhouse

Tunnel Garden

Raised Bed Garden

Living Wall

Choose one part from one of the garden systems.
Write or **draw** what the part is and what its function is.

For the part you chose, why did the designer choose that material?

1.3 Designing for Plants

1.3 | Learn How can humans help plants grow?

What Will You Do?

Now, you will use the materials you have to build a chair. First, **look** at your materials.

Choose three of the materials. What properties do the materials have?

Look at all the materials you have. Which will help you build a chair?

Make a plan for building your chair.

Draw how you will build the chair. **Label** the drawing with the materials you will use.

Now, you and your group can **build** your chair!

1.3 Designing for Plants

1.3 | Learn — How can humans help plants grow?

Think About the Activity

Fill in the T-Chart with what you learned from building the chair. **Write** the materials and their properties on the left side. **Write** how the materials are used on the right side.

Materials and Their Properties	How the Materials Are Used

What changes did you have to make to your plan when you built the real chair?

Do you think most objects are made from one material only, or from many materials joined together? Give some examples.

1.3 | Learn — How can humans help plants grow?

Activity 13

Evaluate Like a Scientist

Quick Code: us1063s

Building a Greenhouse

Gina is helping her teacher build a greenhouse in the school yard.

Which materials should they use to build the different parts of the greenhouse?

Draw arrows to show where each material should be used. The materials may be used in more than one part.

SEP Constructing Explanations and Designing Solutions
CCC Structure and Function

| Glass | Plastic | Wood |

1.3 Designing for Plants | 151

1.3 | Learn How can humans help plants grow?

Activity 14

Solve Problems like a Scientist

Quick Code: us1064s

Designing a Vegetable Garden

In this project, you will draw two designs of a vegetable garden. Your class will make a list of things that a vegetable garden in a city needs. Then, you will use that list to choose the best design.

Vegetables help your body grow. They help you do everything you need to do every day. They help you read, write, play, and even jump!

You can get fresh vegetables from grocery stores. But some neighborhoods, especially in cities, do not have grocery stores that sell fresh vegetables.

What can people who live in these places do to get vegetables? They can grow a vegetable garden. What would a vegetable garden in the city look like? What does it need?

What kind of vegetables should be planted in this garden?

- **SEP** Developing and Using Models
- **SEP** Constructing Explanations and Designing Solutions
- **SEP** Obtaining, Evaluating, and Communicating Information

Think about making a vegetable garden in the city. Where would you put the garden?

A Vegetable Garden

1.3 Designing for Plants | 153

1.3 | Learn How can humans help plants grow?

Make a list of things that a vegetable garden should have. **Discuss** the list as a class. **Add** ideas from the class list to your list.

Draw your designs. Be sure to include these things in your drawings:

- Ways to provide the plants in the garden with the things they need: sunlight, soil, and water

- Labels for the parts of the garden

- Descriptions of what materials you will use to build two parts of the garden

- Names of two vegetables you would plant in the garden

- Pictures of young and adult vegetables

Draw your first vegetable garden design.

1.3 Designing for Plants | 155

1.3 | Learn — How can humans help plants grow?

Draw your second vegetable garden design.

Use the list your class made of things a vegetable garden needs. Which of your two designs is better? How do you know which design is the best?

1.3 Designing for Plants

1.3 | Share How can humans help plants grow?

Activity 15
Record Evidence Like a Scientist

Quick Code: us1065s

Creepers

Now that you have learned about designing for plants, look again at the picture of the Creepers. You first saw this in Wonder.

Let's Investigate Creepers

Talk Together

How can you describe Creepers now? How is your explanation different from before?

SEP Constructing Explanations and Designing Solutions

Look at the Can You Explain? question. You first read this question at the beginning of the lesson.

> **Can You Explain?**
> How can humans help plants grow?

Now, you will use your new ideas about creepers to answer a question.

1. **Choose** a question. You can use the Can You Explain? question, or one of your own. You can also use one of the questions that you wrote at the beginning of the lesson.

   Your Question

2. Then, **use** the sentence starters to answer the question.

1.3 Designing for Plants | 159

1.3 | Share — How can humans help plants grow?

I know that humans help plants grow because

Based on my observations, I think

because

An example of humans helping plants grow would be

An example of how humans use materials to do the same function of a plant part is

The evidence I collected shows

STEM in Action

Quick Code: us1066s

Activity 16
Analyze Like a Scientist

Investigating Soil

Read about investigating soil. Then, **answer** the question.

Read Together

Investigating Soil

What kind of soil is under your feet? You know that there are different kinds of soil. Some plants may grow well in a certain kind of soil.

SEP Obtaining, Evaluating, and Communicating Information

The kids in the video want to find the best place to grow zinnias. Watch how they investigate different kinds of soil.

What Is Dirt?

Did you know that you can get paid to investigate soil? You can be a soil scientist. Soil scientists can tell farmers whether crops will grow well in a certain place. They can tell where wind and water might carry too much soil away. Some of these scientists can even tell what land was like thousands of years ago!

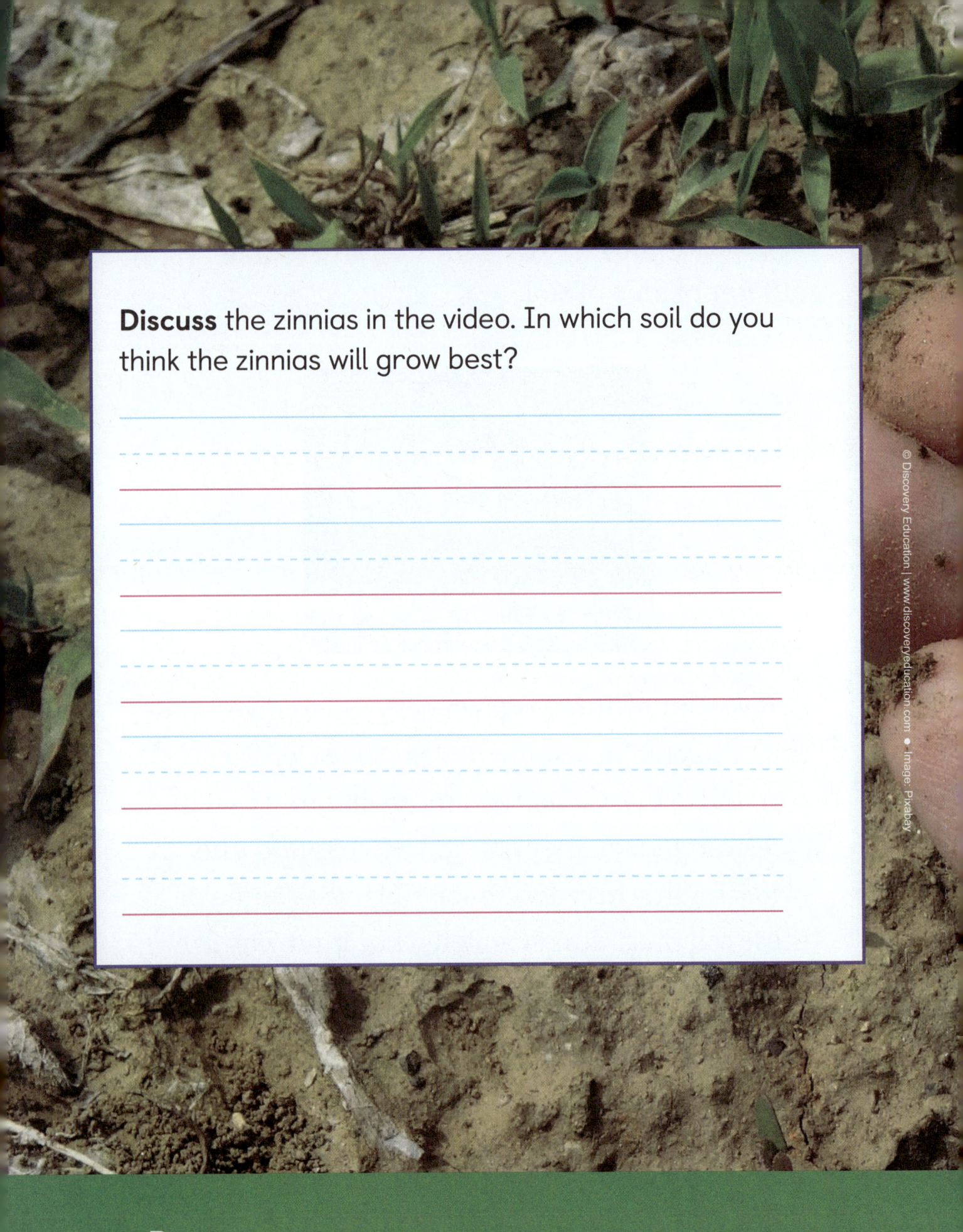

Discuss the zinnias in the video. In which soil do you think the zinnias will grow best?

Which Vegetable?

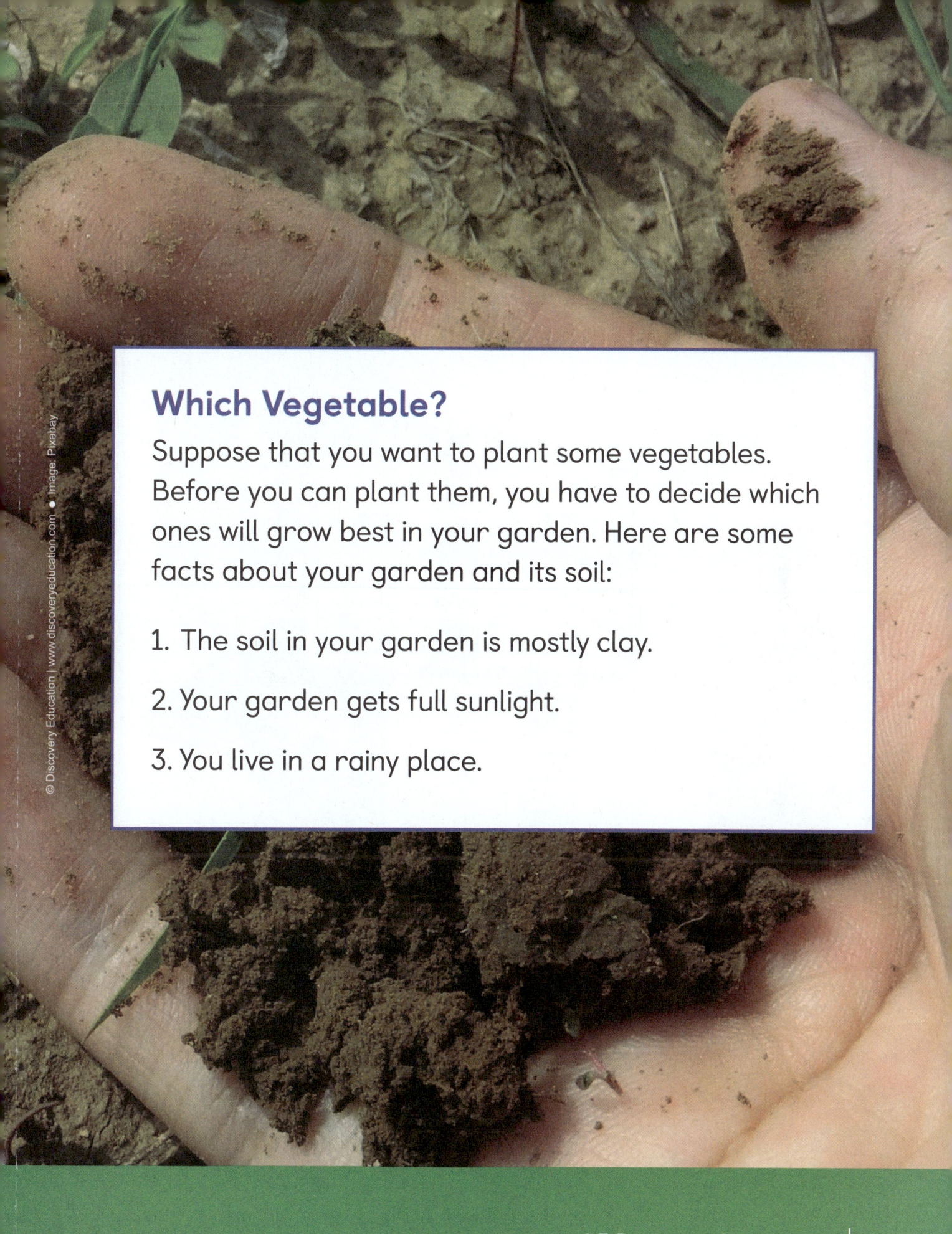

Suppose that you want to plant some vegetables. Before you can plant them, you have to decide which ones will grow best in your garden. Here are some facts about your garden and its soil:

1. The soil in your garden is mostly clay.

2. Your garden gets full sunlight.

3. You live in a rainy place.

Read the seed packets. **Circle** the ones you think will grow best in your garden.

- Carrots: Sandy soil, Full sun, Weekly watering
- Lettuce: Clay soil, Full sun, Daily watering
- Tomatoes: Loam soil, Full sun, Daily watering
- Snap Beans: Clay soil, Full sun, Daily watering
- Zucchini: Sandy soil, Part shade, Weekly watering

Activity 17

Evaluate Like a Scientist

Quick Code: us1067s

Review: Designing for Plants

Think about what you have read and seen. What did you learn?

Draw what you have learned. Then, **tell** someone else about what you learned.

Talk Together

Think about what you saw in Get Started. Use your new ideas to discuss how you can design a garden to grow vegetables.

SEP Obtaining, Evaluating, and Communicating Information

Unit Project

Design Solutions Like a Scientist

Quick Code: us1068s

Hands-On Engineering: Pizza Garden

In this activity, you will design and plan for a garden that can grow toppings for pizza.

Tomato Plant

- **SEP** Asking Questions and Defining Problems
- **SEP** Developing and Using Models
- **SEP** Constructing Explanations and Designing Solutions
- **SEP** Analyzing and Interpreting Data
- **CCC** Scale, Proportion, and Quantity
- **CCC** Structure and Function

What materials do you need?
(per group)

- Pictures of adult and young plants
- Metric ruler
- Paper

Ask Questions About the Problem

Choose two or more plants to grow in your garden.

Think about what your garden will look like.
Write your ideas here.

Unit 1: Plant Shapes | 169

Unit Project

How can you be sure the plants will get what they need to survive?

What Will You Do?

Draw a picture of your design.

Test your design. **Draw** or **write** to show how you tested it.

Think About the Activity

Write or **draw** your answers to the questions in the chart.
How well did your design help your plants grow?
How could you improve your design?

What Worked?	What Didn't Work?

What Could Work Better?

Unit 1: Plant Shapes | 171

Grade 1 Resources

- **Bubble Map**
- **Safety in the Science Classroom**
- **Vocabulary Flash Cards**
- **Glossary**
- **Index**

Name _____

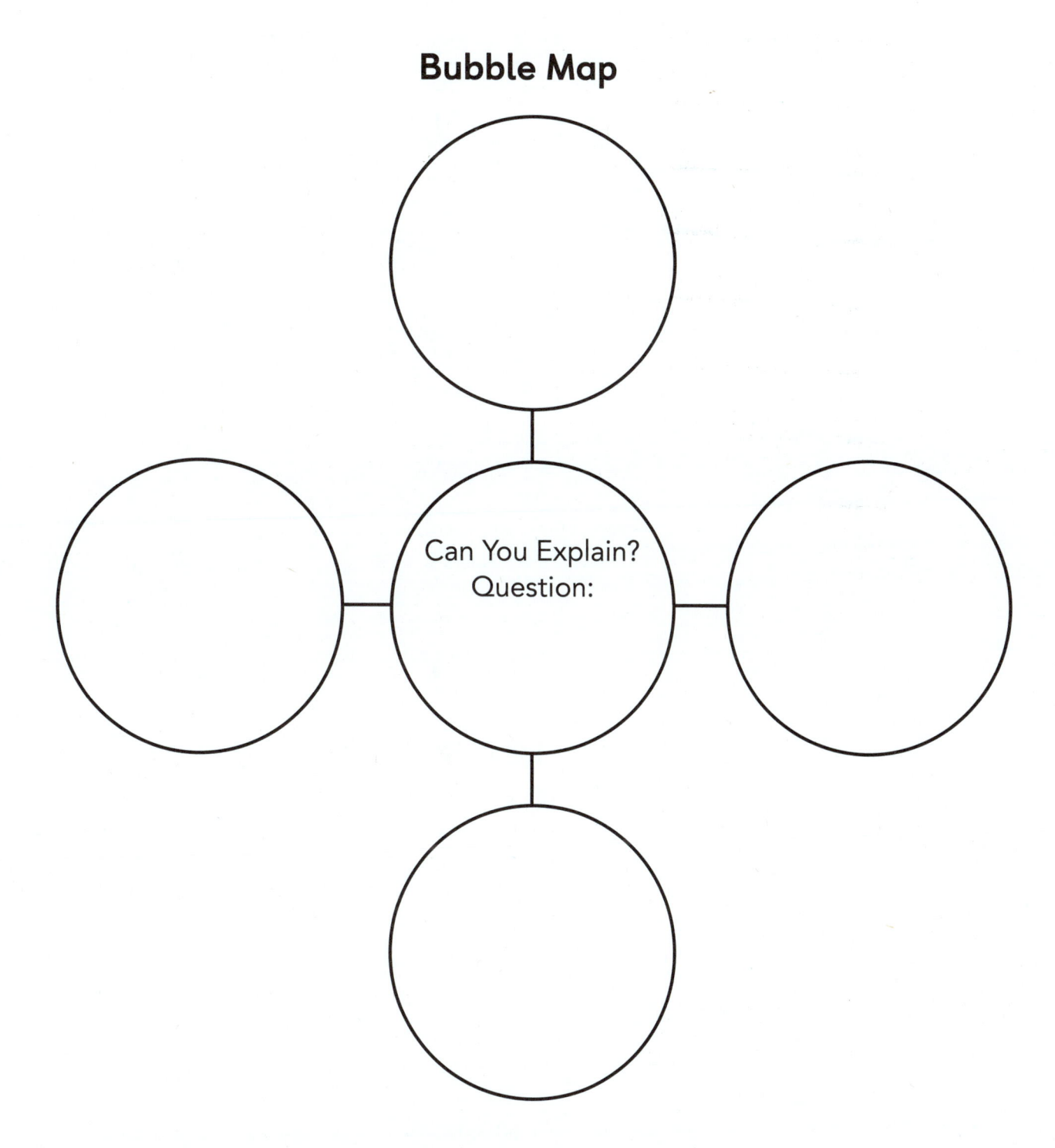

Bubble Map

Can You Explain?
Question:

Bubble Map | R1

Safety in the Science Classroom

Following common safety practices is the first rule of any laboratory or field scientific investigation.

Dress for Safety

One of the most important steps in a safe investigation is dressing appropriately.

- Splash goggles need to be kept on during the entire investigation.

- Use gloves to protect your hands when handling chemicals or organisms.

- Tie back long hair to prevent it from coming in contact with chemicals or a heat source.

- Wear proper clothing and clothing protection. Roll up long sleeves, and if they are available, wear a lab coat or apron over your clothes. Always wear closed-toe shoes. During field investigations, wear long pants and long sleeves.

Safety Goggles

Be Prepared for Accidents

Even if you are practicing safe behavior during an investigation, accidents can happen. Learn the emergency equipment location in your classroom and how to use it.

- The eye and face wash station can help if a harmful substance or foreign object gets into your eyes or onto your face.

- Fire blankets and fire extinguishers can be used to smother and put out fires in the laboratory. Talk to your teacher about fire safety in the lab. He or she may not want you to directly handle the fire blanket and fire extinguisher. However, you should still know where these items are in case the teacher asks you to retrieve them.

Most importantly, when an accident occurs, immediately alert your teacher and classmates. Do not try to keep the accident a secret or respond to it by yourself. Your teacher and classmates can help you.

Fire Extinguisher

Practice Safe Behavior

There are many ways to stay safe during a scientific investigation. You should always use safe and appropriate behavior before, during, and after your investigation.

- Read all of the steps of the procedure before beginning your investigation. Make sure you understand all the steps. Ask your teacher for help if you do not understand any part of the procedure.

- Gather all your materials and keep your workstation neat and organized. Label any chemicals you are using.

- During the investigation, be sure to follow the steps of the procedure exactly. Use only directions and materials that have been approved by your teacher.

- Eating and drinking are not allowed during an investigation. If asked to observe the odor of a substance, do so using the correct procedure known as wafting, in which you cup your hand over the container holding the substance and gently wave enough air toward your face to make sense of the smell.

- When performing investigations, stay focused on the steps of the procedure and your behavior during the investigation. During investigations, there are many materials and equipment that can cause injuries.

- Treat animals and plants with respect during an investigation.

- After the investigation is over, appropriately dispose of any chemicals or other materials that you have used. Ask your teacher if you are unsure of how to dispose of anything.

- Make sure that you have returned any extra materials and pieces of equipment to the correct storage space.

- Leave your workstation clean and neat. Wash your hands thoroughly.

Vocabulary Flash Cards

absorb
to take in or to soak up

edible
able to be eaten as a food

energy
the ability to do work or make something change

engineer
a person who designs something that may be helpful to solve a problem

Vocabulary Flash Cards | R7

flower

the plant part that blooms with colorful petals and beautiful smells and holds the part of the plant that makes the seeds

fruit

the plant part that contains seeds and grows from a flowering plant

leaf

the part of the plant that grows off the stem and collects sunlight for the plant to make food

light

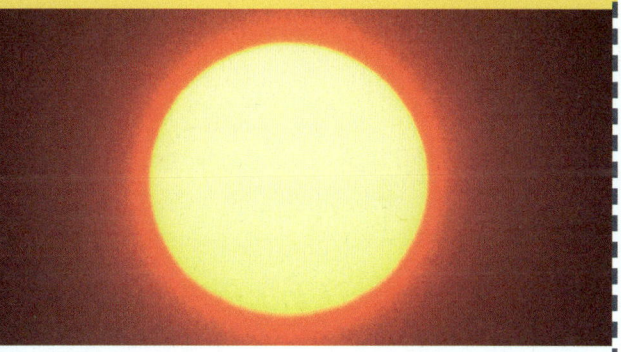

a form of energy that makes it possible for your eyes to see

Vocabulary Flash Cards | **R9**

material

things that can be used to build or create something

nutrient

something in food that helps people, animals, and plants live and grow

plant

a living thing made up of cells that needs water and sunlight to survive

seed

the small part of a flowering plant that grows into a new plant

Vocabulary Flash Cards | R11

seedling

a baby plant that starts from a seed

soil

dirt that covers Earth, in which plants can grow and insects can live

stem

the part of the plant that grows up from the roots and holds up the leaves and flowers

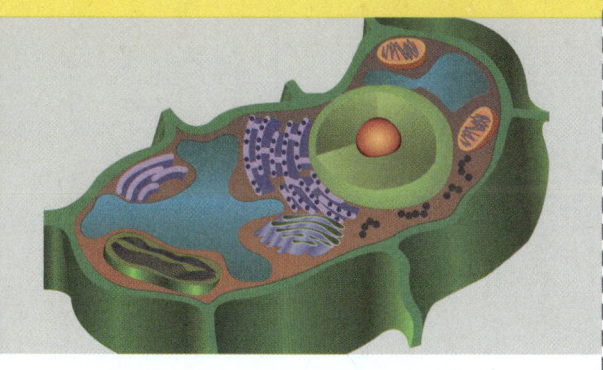

structure

a part of an organism; the way parts are put together

Vocabulary Flash Cards | R13

system

a group of parts that go together to make something work

tendril

a long, thin stem that wraps around things as it grows

trait

a characteristic that you get from one of your parents

water

a clear liquid that has no taste or smell

Vocabulary Flash Cards | R15

Glossary

English ——— A ——— Español

absorb
to take in or soak up

absorber
tomar o captar

air
an invisible gas that is all around us; living things, such as plants and animals, need it to breathe and grow

aire
gas invisible que nos rodea; todos los seres vivos, como las plantas y los animales, lo necesitan para respirar y crecer

animal
a living thing that moves around to look for food, water, or shelter, but can't make its own food

animal
ser vivo que se mueve para buscar alimento, agua o refugio, pero no puede producir su propio alimento

axis
a real or imaginary line through the center of an object; the object turns around it

eje
línea real o imaginaria que pasa por el centro de un objeto; el objeto gira alrededor de ella

B

binoculars
a device that is put up to your eyes so you can see far away

binoculares
dispositivo que se pone sobre los ojos para poder ver lejos

C

collect
to gather

recolectar
reunir

communicate
to give and get information, messages, or ideas

comunicarse
dar y recibir información, mensajes o ideas

constellation
a particular area of the sky; a group of stars

constelación
área particular del cielo; grupo de estrellas

E

Earth
the third planet from the Sun; the planet on which we live (related words: earthly; earth - meaning soil or dirt)

Tierra
tercer planeta desde el Sol; planeta en el cual vivimos (palabras relacionadas: terrenal; tierra en el sentido de suelo o barro)

edible
able to be eaten as a food

comestible
que se puede comer como alimento

energy
the ability to do work or make something change

energía
habilidad de trabajar o producir un cambio

engineer
a person who designs something that may be helpful to solve a problem

ingeniero
persona que diseña algo que puede ser útil para resolver un problema

F

feature
a thing that describes what something looks like; part of something

rasgo
cosa que describe cómo se ve algo; parte de algo

flower
the plant part that blooms with colorful petals and beautiful smells and holds the part of the plant that makes the seeds

flor
parte de la planta que florece con pétalos de colores y aromas agradables y contiene la parte que produce las semillas

fruit
the plant part that contains seeds and grows from a flowering plant

fruta
parte de la planta que contiene semillas y crece de una planta en flor

I

inherit
to receive a characteristic from one's parents

heredar
recibir una característica de los padres de alguien

L

leaf
the part of the plant that grows off the stem and collects sunlight for the plant to make food

hoja
parte de la planta que crece desde el tallo y reúne luz solar para que la planta produzca alimento

light
a form of energy that makes it possible for our eyes to see

luz
forma de energía que hace posible ver con los ojos

M

material
things that can be used to build or create something

material
cosas que se pueden usar para construir o crear algo

measure
to find the amount, the weight, or the size of something

medir
hallar la cantidad, el peso o el tamaño de algo

moon
any object that goes around a planet

luna
cualquier objeto que gira alrededor de un planeta

N

nutrient
something in food that helps people, animals, and plants live and grow

nutriente
algo en los alimentos que ayuda a las personas, los animales y las plantas a vivir y crecer

O

observe
to watch closely

observar
mirar atentamente

opaque
when no light gets through something, like wood or metal

opaco
cuando no pasa luz a través de algo, como la madera o el metal

orbit
to travel in a circular path around something

orbitar
viajar un recorrido circular alrededor de algo

P

plant
a living thing made up of cells that needs water and sunlight to survive

planta
ser vivo formado por células que necesita agua y luz solar para sobrevivir

position
a place where a person or a thing is located

posición
lugar donde se encuentra una persona o cosa

property
a characteristic of something

propiedad
característica de algo

--- **R** ---

reflect
when something like light or heat bounces off a surface

reflejar
cuando algo como la luz o el calor rebota en una superficie

rotate
to turn around a center point; to spin

rotar
girar alrededor de un punto central; dar vueltas

--- **S** ---

seed
the small part of a flowering plant that grows into a new plant

semilla
parte pequeña de una planta en flor que se convierte en una nueva planta

seedling
a baby plant that starts from a seed

plántula
planta joven que crece de una semilla

sense
sight, hearing, smell, taste, or touch

sentido
visión, audición, olfato, gusto o tacto

soil
dirt that covers Earth, in which plants can grow and insects can live

suelo
tierra que cubre nuestro planeta en la que pueden crecer plantas y vivir insectos

sound
anything that people or animals can hear with their ears

sonido
todo lo que las personas o los animales pueden oír con los oídos

source
the start or the cause of something

fuente
el comienzo o la causa de algo

star
a burning ball of gas in space

estrella
bola ardiente de gas en el espacio

stem
the part of the plant that grows up from the roots and holds up the leaves and flowers

tallo
parte de la planta que crece hacia arriba desde las raíces y sostiene las hojas y las flores

structure
a part of an organism; the way parts are put together

estructura
parte de un organismo; la forma en que se unen las partes

sun
any star around which planets revolve

sol
toda estrella alrededor de la cual giran los planetas

system
a group of parts that go together to make something work

sistema
grupo de partes que se combinan para hacer que algo funcione

T

technology
inventions that were developed to solve problems and make things easier

tecnología
inventos que se desarrollaron para resolver problemas y hacer más fáciles las cosas

tendril
a long, thin stem that wraps around things as it grows

zarcillo
tallo largo y delgado que se enrosca alrededor de cosas a medida que crece

trait
a characteristic that you get from one of your parents

rasgo
característica que se recibe de uno de los padres

translucent
when some light gets through and what is on the other side might not be very clear, like fog

translúcido
cuando pasa algo de luz y lo que hay del otro lado puede no ser muy transparente, como la niebla

transparent
when light passes through and you can see clearly, such as clean water and air

transparente
cuando pasa la luz y se puede ver con claridad, como el agua limpia y el aire

--- V ---

vibration
the rapid movement of an object back and forth

vibración
rápido movimiento de un objeto adelante y atrás

volume
the loudness of a sound

volumen
la intensidad de un sonido

--- W ---

water
a clear liquid that has no taste or smell

agua
líquido transparente que no tiene sabor ni olor

wave
the way sound moves through the air

onda
manera en la que el sonido viaja por el aire

Index

A

Absorbing
   in greenhouses, 139
   by plants, 30
Air, growing plants in, 105
Airports, 140
Analyze Like a Scientist,
   15–16, 30–33, 41, 43, 45,
   52–55, 83–85, 88–89,
   104–107, 118–120,
   138–140, 162–166
Animals
   needs of, 67
   in observation area, 66
Architects, 138
Ask Questions Like a
   Scientist, 10–11, 62–63,
   114–117

B

Banyan fig tree, 118
Bees, 84
Beetles, 119
Biomimicry, 118–120
Butterfly gardens, 17, 18

C

Can You Explain?, 8, 60, 112
Chairs, 146–149
Cities, gardens in, 106,
   152–157
Comparing
   needs of plants and
     animals, 67
   plants, 23, 28, 34–38,
     92–96
   seed growth, 77
   shadows of different
     leaves, 95
   shapes of plant parts, 136
Creeper plants, 114–115,
   158–161

D

Desert plants, 34
Design
   of chairs, 147, 149
   for gardens, 130–131
   model for, 123
   from nature, 120
   for plants, 121–123, 167

Design (*cont.*)
   for solving problems, 122
   of structures, 138–140
   for vegetable garden, 152–157
Design Solutions Like a Scientist, 4–5, 168–171

E

Edibles, 40
Engineers, 138
Environment, studying, 64–68
Evaluate Like a Scientist, 17, 42, 46–47, 56, 90–91, 99, 108, 121–123, 141, 150–151, 167

F

Ferns, 140
Flowers, 14
   and bees, 84
   different and similar, 15
   different kinds of, 21, 23
   edible, 40, 42
   of old plants, 44

Food deserts, 2
Foods
   introducing new, 52–55
   plants making their own, 30
Fruits
   edible, 40, 42
   introducing new, 52–53
   of old plants, 44
   in salads, 11, 16
   and seeds, 39
   vegetables and, 35

G

Gardens
   butterfly, 17, 18
   in cities, 106, 152–157
   different designs for, 130–131
   growing, 124–129
   pizza, 4–5, 168–171
   planting, 132
   urban, 2
   vegetable, 152–157

Garden salads, 10–11, 15–16, 43, 48–51
Garden systems, 144–145
Gotham Greens, 106
Graphs, 38
Greenhouses
    building, 150–151
    garden systems in, 139
Growing plants, 60, 108
    different methods of, 107
    in gardens, 124–129
    humans helping, 112, 160–161
    and plant parts, 71, 88–89
    from seeds, 74–82
    without soil, 104–107

H

Hydroponic plants, 62–63, 100–103, 104–107

I

Investigate Like a Scientist, 74–81, 92–96, 142–149
Irrigation systems, 139

L

Leaves
    of banyan fig tree, 118
    characteristics of, 30
    comparing, 92–96
    different kinds of, 22, 23
    edible, 40, 42
    and growing plants, 83
    in growing plants, 89
Light
    absorbing, 139
    for growing plants indoors, 105
Living things, needs of, 70
Locations, for gardens, 132

M

Mangrove, 140
Materials
    for building chairs, 146, 148
    for building greenhouses, 150–151
    properties of, 142–149
    used in building objects, 149
Model for design, 123

N

Nature, copying designs from, 120
Non-edible plants, 41
Nutrients, for plants, 83

O

Objects, in observation area, 66
Observation area, 64–68
Observe Like a Scientist, 12–13, 14, 25–27, 28–29, 34–38, 39, 69–70, 86–87, 97–98, 124–126, 127–129, 130–131, 132
Old plants, 44–47
Outdoors, exploring, 20–24

P

Parts
   of plants (*See* Plant parts)
   of structures, 141
   working together, 121
Pizza garden, 4–5, 168–171
Plant cycles, 18
Plant parts
   describing, 19, 33
   edible, 40
   functions of, 71, 86–87, 90–91, 134
   helping plants grow, 88–89, 103
   labeling, 25–27
   and needs of plants, 85
   shapes of, 135
Plants
   changes in, 8, 17, 50
   characteristics of, 30–33
   comparing, 34–38, 92–96
   edible, 40–42
   growing (*See* Growing plants)
   growing without soil, 104–107
   needs of, 26, 67, 69–70, 72–73, 83–85
   non-edible, 41
   in observation area, 66
   observations about, 20–24

in salads, 15–16, 43
shapes of, 133–137
structures of, 114–117
and sunlight, 97–98, 99
watering, 28–29
young and old, 44–47
Properties, of materials, 142–149
Puri, Viraj, 106

Q

Questions, asking, 68

R

Record Evidence Like a Scientist, 48–51, 100–103, 158–161
Red peppers, 12–13
Roof
    design of, 118
    in greenhouse, 139
Roots
    characteristics of, 30, 31
    edible, 40, 42
    in growing plants, 89

S

Seeds
    different kinds of, 22, 23
    edible, 40, 42
    and fruits, 39
    growing plants from, 44, 74–82, 88
    needs of, 25
Senses, in exploring, 68
Shadows, 92–96
Shapes
    of plant parts, 135
    of plants, 133–137
Soil
    growing plants without, 104–107
    investigating, 162–166
    roots in, 30, 31
    for vegetable garden, 154
Soil scientists, 163
Solve Problems Like a Scientist, 152–157
Sorting plants, 29

Stems
    characteristics of, 31
    edible, 40
    in growing plants, 89
Structures
    designing, 138–140
    parts of, 139, 141
    of plants, 114–117
Sunlight
    and plants, 30, 84, 97–99
    and shadows, 92–96
    for vegetable garden, 154
Systems, 139

T

Tents, 121
Think Like a Scientist, 20–24, 64–68, 133–137
Train stations, 140
Traits, of plants, 37
Trees
    designing structures like, 138
    different kinds of, 22, 23

U

Urban gardens, 2

V

Variables, in growing plants, 125
Vegetables
    designing a garden for, 152–157
    fruits and, 35
    growing, 165–166
    introducing new, 52–53
    in salads, 11
Venus flytrap, 116

W

Water
    collecting, 119
    growing plants in, 104–107
    for plants, 28–29
    sliding off leaves, 118
    for vegetable garden, 154

Y

Young plants, 44, 45, 46–47

Z

Zinnias, 163–164